Epistemic Processes

Inge S. Helland

Epistemic Processes

A Basis for Statistics and Quantum Theory

 Springer

Inge S. Helland
Department of Mathematics
University of Oslo
Oslo, Norway

ISBN 978-3-030-06968-1   ISBN 978-3-319-95068-6   (eBook)
https://doi.org/10.1007/978-3-319-95068-6

This Springer imprint is published by the registered company Springer Nature Switzerland AG
The registered company address is: Gewerbestrasse 11, 6330 Cham, Switzerland

Preface

All human decisions, including those made in scientific experiments and in observational studies, are made in some context. Such contexts are being explicitly considered in this book. To do so, a conceptual variable is defined as any variable which can be defined by a (group of) human(s)—for instance scientists—in a given setting. Such variables are classified. Sufficiency and ancillarity in a statistical model used in scientific investigations are defined conditionally in the scientists' context. The conditionality principle, the sufficiency principle and the likelihood principle are generalized, and a rule for when one should not condition on an ancillary is motivated by examples. Model reduction is discussed in general from the point of view that there exists a mathematical group acting upon the parameter space of the model. It is shown that a natural extension of this whole discussion gives a conceptual fundament from which the formalism of quantum theory can be discussed. This can also be considered as an argument for an epistemological basis for quantum theory, a kind of basis that has also been advocated by part of the quantum foundation community in recent years. Born's celebrated formula is shown to follow from a focused version of the likelihood principle together with some reasonable assumptions on rationality connected to experimental evidence. The questions around Bell's inequality are approached by using an epistemic point of view connected to each observer. The Schrödinger equation is derived from reasonable assumptions. The objective aspects of the world are identified with the ideal observational results upon which all real and imagined observers agree. Philosophical conclusions from my point of departure are discussed.

This is a very brief summary of this book. At the outset, it is written for several groups of readers: (a) physicists interested in the foundation of their science, (b) statisticians interested in the foundation of their science and its relationship to modern physics, (c) philosophers of science, (d) students with a good background in mathematics and (e) mathematicians with an interest in the foundation of empirical science. To have such a diversity of readers in mind is quite a challenge. Some will experience part of the text as rather technical; others will see the approach as unusual compared to the way they are used to see their own science presented. I encourage the reader to start with a relatively open mind, but also criticism of my way of thinking is welcome. The text concentrates on the foundation; there are very many aspects of both quantum theory and statistical inference that are not covered. On the foundational level, I argue that there is a connection between these two areas. This may be an unusual way of thinking, especially for physicists, and it implies a special interpretation of quantum mechanics. As an interpretation it is related to the recent Quantum Bayesian approach, briefly called QBism, but I try to emphasize that Bayesianism is a branch of statistical inference theory, not of probability theory. Also, I will allow other approaches to statistical inference. Looking at quantum mechanics as related to statistical inference will presuppose that there are data. These data are provided by macroscopic measurement apparatuses. In the ordinary approach towards quantum mechanics, these measurement apparatuses are first discussed at a later stage; here I will assume from the beginning that they are there.

In a discussion of a brief version of the account given in this book, the Quantum Bayesian Ruediger Schack wrote: "The main difference between your approach and QBism is that your approach is phrased in terms of acquiring information about variables. QBism on the other hand is concerned with decision theory...". I will discuss my relation to QBism in more detail below. I appreciate that decision theory is important, but here I prefer the Quantum Decision Theory promoted in a series of papers by Yukalov and Sornette; see Sect. 1.3 below. Also, see my remarks to this theory in Sect. 5.5, where I just see *decision* as a primitive concept. My claim is that a theory about acquiring information about variables is important enough to warrant a book. After all its aim is to cover the basis of statistics as a science and also very much of quantum theory as a science. Some of my more qualitative views on the process of making decisions are discussed in Chap. 6.

Statisticians talk about data on the one hand and parameters of statistical models on the other hand. Here, I will generalize the parameter concept in such a way that the new concept also covers prediction and other statements about single units. I call the new concept *e-variable*, an abbreviation for epistemic conceptual variable. In a quantum mechanical setting, simple e-variables will correspond to what is usually is called observables. As a statistician I want to distinguish between these observables and the corresponding data values that result from experiments. A statistician will call these estimates. However, in quantum theory, this distinction is not always clear. The reason from a statistical point of view is as follows: *The e-variables of elementary quantum theory are discrete, and experiments are often very accurate. This result, in confidence intervals, credibility intervals or prediction intervals around the true value that may degenerate to a single point, which actually is the true value. Thus, in such cases the distinction between estimates and true value is blurred out, and much of statistical theory becomes irrelevant.* At the foundational level, I will nevertheless keep the distinction and insist that this is related to statistical inference. As such, it is made in a context, as already mentioned. The observer, and the information available to the observer, plays a crucial role in quantum mechanics.

I acknowledge that many readers may be confused by this book, especially readers who know a little (or more) of the usual treatment of quantum mechanics. I think that the only thing I can offer these readers at present is a letter that I wrote to my colleague Barbara Heller recently. Barbara is a statistician living in Chicago, and she has a keen interest in quantum mechanics. After reading a part of my manuscript, she wrote to me that she was confused. Here is my answer.

Dear Barbara,

I understand very well your confusion. —

What does statistics have to offer, in my opinion? It is true that statistics is being used to a large extent by experimental physicists, but I have been looking for something more fundamental. Can one think of quantum theory and statistical inference as having partly a common basis? I think so, and I have tried to sketch such a basis in my book. The common basis is, as I see it, that of an epistemic process: a process to achieve knowledge. In the simplest case having what I call an e-variable (epistemic conceptual variable) θ, asking a question to nature: what is θ? And obtaining an answer in terms of information about θ.

In statistics, θ is most often a continuous parameter, and the answer can be in terms of a confidence interval or a Bayesian credibility interval. In elementary quantum mechanics, θ can be discrete, and in an ideal measurement we can obtain a definite answer: $\theta = u$.

Furthermore, in quantum theory, we have what I call complementary e-variables: think of spin components in different directions. We can obtain information about θ^a and about θ^b, but the vector (θ^a, θ^b) is inaccessible.

In the Hilbert space formulation, an e-variable is identified with an operator, the possible values of θ are the eigenvalues of that operator and very many state vectors can be identified by the events of the type $\theta = u$. I have made a point of discussing this identification from several points of view in my book.

Where do the probabilities in quantum theory come from? They are given by Born's formula, a formula for $P(\theta^b = u_k | \theta^a = u_j)$ in terms of the state vectors. I have a rather long derivation of Born's formula in my book, using two premises: (1) A focused likelihood principle, which can be motivated from the likelihood principle in statistics, a basic principle which again can be derived from fairly obvious assumptions: a conditionality principle and a sufficiency principle. (2) An assumption of perfect rationality.

Of course, this is not the full story, but it appears to me to be a very useful beginning for understanding quantum theory from assumptions that are related to the basic assumptions of statistical theory.

I hope that this has clarified my points of view to some extent, and I sincerely hope that it may have contributed to remove some of your confusion.

*

The readers just interested in my views on the foundation of quantum mechanics may skip Chap. 2, but Sect. 2.3 should be looked at, and the concept of likelihood as the joint probability (probability density) of the data, seen as a function of the e-variable/parameter (Sect. 2.1.3), should be understood. Section 3.2 may be dropped if the generalized likelihood principle of Sect. 3.2.4 is taken for granted. Section 4.4 may be dropped in the first reading. The very impatient reader may go directly from Chap. 1 to Sect. 5.15.

Aas, Norway   Inge S. Helland

References

Bargmann, V. (1964). Note on Wigner's Theorem on symmetry operations. *Journal of Mathematical Physics, 5*, 862–868.

Bjørnstad, J. F. (1990). Predictive likelihood: A review. *Statistical Science, 5*, 242–265.

Gelman, A., & Robert, C. P. (2013). "Not only defended but also applied": The perceived absurdity of Bayesian inference. *The American Statistician, 67*, 1–5.

Hammond, P. J. (2011). *Laboratory Games and Quantum Behavior. The Normal Form with a Separable State Space*. Working paper. Department of Economics, University of Warwick.

Helland, I. S. (2008). Quantum mechanics from focusing and symmetry. *Foundations of Physics, 38*, 818–842.

Smolin, L. (2011). A real ensemble interpretation of quantum mechanics. aXiv. 1104.2822 [quant-ph].

Wigner, E, P. (1959). *Group theory and its application to the quantum mechanics of atomic spectra*. New York: Academic.

Acknowledgements

I am grateful to Philip Goyal for inviting me to the workshop on Reconstructing Quantum Theory at the Perimeter Institute in 2009 on the basis of Helland (2008). Gudmund Hermansen has done some of the calculations in connection to Example 5.16. Also, thanks to Arne B. Sletsjøe for discussions in effect leading to Proposition 4.2, to Erik Alfsen for giving me the preprint Hammond (2011), to Kingsley Jones for making me aware of Wigner (1959) and of Bargmann (1964), to Joan Ennis for clarification of the meaning of the term "epistemic", to Chris Ennis for sending me Smolin (2011), for recommending to me the book by Anton Zeilinger and for general discussions, to Pekka Lahti for information related to the Born rule, to Dag Normann for information about propositional logic, to Gudmund Hermansen, Celine Cunen and Emil Aas Stoltenberg for the reference to Gelman and Robert (2013), to Nils Lid Hjort for making me aware of Bjørnstad (1990), to Barbara Heller for giving her first reactions to an unfinished manuscript, to Bent Selchau for reading and giving comments to Chap. 1 and not least for helping me through my own epistemic processes, to Paul Busch for comments on the final manuscript and finally to Harald Martens and Chris Fuchs for discussions. Special thanks go to Andrei Khrennikov for throughout the years arranging very interesting conferences on quantum foundation. Big thanks go to Sharon Tubb and the rest of the Francisco Street gang for their emotional and intellectual stimulations whenever we meet. Finally, I want to thank my family for everything they have sacrificed while I have been seeking a foundation related to my views of science.

Contents

Chapter 1
The Epistemic View Upon Science

Abstract This chapter gives the background for the book. Its relation to other views on the foundation of quantum theory are clarified and discussed. The fundamental notion of an e-variable (epistemic conceptual variable) is explained, it is discussed and is related to the statistical parameter-concept. A quantum state is in some generality linked to a question-and-answer pair, and an experiment connected to such a question-and-answer pair is described for the case of a spin 1/2 particle. The two basic postulates of quantum theory are stated and discussed. The importance of inaccessible conceptual variables is stressed, and this is related to Bohr complementarity.

1.1 Introduction

The aim of science is to gain knowledge about the external world; this is what we mean by an epistemic process. In its most primitive form, the process of achieving knowledge can be described by what Brody (1993) called an epistemic cycle: "Act, and see what happens". Experiments in laboratories and observational studies done by scientists are usually much more sophisticated than this; they often require several epistemic cycles and also higher order epistemic cycles acting upon the first order cycles. An experiment or an observational study is always focused on some concrete system, it involves concrete experimental/observational questions and it is always done in a context, which might depend on conceptual formulations; in addition the context may be partly historical and partly chosen by the scientist himself, or depending upon the scientist.

In earlier years, experiments were often done by single scientists; now it is more and more common that people are working in teams. Also, results of experiments should be communicated to many people. This calls for a conceptual basis which is common to a whole culture of scientists. One problem, however, is that people from different scientific cultures have difficulties with communicating. They might not have a common language. The first purpose of this book is to develop a scientific language for achieving knowledge which is a synthesis of the languages that I have met in the three cultures that have been exposed to myself: (1) Mathematical statistics; (2) Quantum mechanics; (3) Applied statistics

© Springer-Verlag GmbH Germany, part of Springer Nature 2018

I. S. Helland, *Epistemic Processes*, https://doi.org/10.1007/978-3-319-95068-6_1

including simple applications. It is a hope that this investigation may lead to a deeper understanding of the epistemic process itself. It is also a hope that such an investigation may be continued in order to include more scientific cultures, say, machine learning and quantum computation.

Since statistical inference is used as a tool in very many experimental studies, also within physics, it is natural to take this culture as a point of departure. But I will add some elements which are not very common in the statistical literature:

1. I make explicit that every experimental investigation is made in a context.
2. A transformation group may be added to the statistical model.
3. Model reductions by means of such groups are introduced.
4. In order to address also the physicists, the parameter concept is introduced through the more general term *epistemic conceptual variable* (e-variable). Any variable which can be defined in words by a person or by a group of persons in an experimental situation is called a conceptual variable. The notion of an e-variable at the outset includes every conceptual variable involved in an epistemic process, and an e-variable can also be connected to a single unit (say to a single human being in a sociological or psychological investigation or to a single particle in physics). The basic aim of an epistemic process is to gain some knowledge about the relevant e-variables. It turns out, however, that even in the basic quantum mechanical situation, the term 'e-variable' might in principle from a statistical point of view be replaced by 'parameter'. The problem is that the latter word is so over-burdened in physics. It is of course important to stress that whenever I say 'e-variable' in connection to an ordinary statistical investigation, this term can be replaced by 'parameter'.
5. To give a conceptual basis for causality theory and ultimately for finding a link to quantum theory, I will also introduce *inaccessible conceptual variables*, that is, conceptual variables which cannot be estimated or given any value with arbitrary accuracy in any experiment. Versions of such unobservable variables can be found in counterfactual situations, but the notion is also relevant, say, in connection to regression models where the number of variables by necessity is larger than the number of units. We will see that this notion is crucial in quantum mechanics, where it can be linked to Niels Bohr's concept of complementarity.

Also, I have included the recent notion of confidence distributions, in order to allow both a frequentist and a Bayesian basis for any given experimental investigation.

This framework as further developed in the present book will lead to a non-formal way to discuss essential elements of quantum theory, a theme which occupies later chapters of this book, and is also discussed further in this chapter. The usual basis for quantum theory as developed by von Neumann (1932) was a great achievement, and the language that is implied by this basis is used for all further theoretical developments and for all discussions among physicists. It is a strong intrinsic part of the quantum mechanical culture, in fact, of the culture shared by the whole community of modern physicists. But since the traditional language is purely formal and has little or no intuitive basis for people outside the community of physicists and mathematicians, it seems to be of some interest to develop an alternative language for discussing aspects of quantum theory, even if this language is somewhat limited compared to the ordinary quantum language.

Several recent investigators in quantum foundations have reasoned that quantum mechanics should be interpreted as an epistemic science. I agree with this. But I see it as somewhat problematic that this notion of an epistemic science should be connected to one language in fundamental physics and a completely different language in the rest of empirical science. One purpose of this book is to argue for a new language concerning the concept of a quantum state. I will keep the notion of a quantum state defined as a unit vector (or ray) in a complex Hilbert space. But in many connections this notion can be replaced by the following: One poses a focused question about the system under consideration: 'What is the value of θ?' and obtains a definite answer: $\theta = u_k$. Here θ is an e-variable/ observable (to be further discussed below), and u_k is one of the values that θ can take. There are open ends of the present programme as far as quantum physics is concerned, but I will argue that the investigations can be carried on further along the same lines.

The sceptic might ask: What is the purpose of introducing a new language when this does not lead to anything new? My first answer is that a simple language may be of importance in communication between people. For those who know this paradox, it may be that Wigner's friend is ignorant of the formal language of quantum theory. Nevertheless, he might have an intuitive feeling of what it means that the spin component in direction a of a particle is $+1$, and from this he might be able to communicate his state notion to Wigner, and the two will then share a common state for the physical system.

My second answer is that I will show that my programme indeed leads to something essentially new, also within the science of quantum mechanics itself: The Born formula, which is the basis for all probability calculations in quantum physics, is taken as an independent axiom in textbooks. I will derive it from a set of intuitive assumptions. I am also able to discuss the problematic questions connected to Bell's inequalities by using an epistemic point of departure. Several so-called paradoxes can be resolved using the language of epistemic processes. Also questions around the derivation of the Schrödinger equation are discussed.

One aspect of my programme is to propose sort of a link between quantum theory and statistical inference, two cultures which until now have been completely separated. The study of scientific cultures is not common. An exception is the book by Knorr Cetina (1999), where the author describes from the inside epistemic cultures connected to two empirical groups: High energy physics experimenters at CERN and molecular biologists at a laboratory. Her arguments strongly depend upon the notion of knowledge societies. Of course I agree that the nature of knowledge is different in different scientific communities, but it is the process of *achieving knowledge* that I feel should have something in common, and it is this process I will focus upon in this book.

As already mentioned, the language for quantum mechanics used in the present book also implies a particular interpretation of the theory. Nearly since its introduction in the beginning of the previous century, the physical community has been divided on the question of the interpretation of quantum theory. On the one hand many has argued for an ontic interpretation of the quantum state: It is a real state of nature. But on the other hand other people has argued for an epistemic interpretation of the quantum state: It only describes an observer's knowledge of the state of nature. In my opinion some sort of a synthesis of the two views is called for. The phenomenon of collapse of the wave packet during a measurement and paradoxes such as that of the Schrödinger cat give strong arguments in order that the epistemic view should play an important part. This in itself calls for a thorough analysis of how the epistemic process can be, and this is part of the purpose of the present book. The observer and his context play an important role in this process. By verbal communication and with the help of time, several observers may develop a common context. The ontic state of a particular physical system is in this book identified with the hypothetical state that all potential ideal observers with a common context of relevance to this physical systems agree upon.

Recently, several related no-go theorems have appeared in the physical literature which have been taken as arguments that a pure epistemic view are inconsistent with the predictions of quantum mechanics. However, these theorems rely on certain specific assumptions. The epistemic toy model of Spekkens (2007), which reproduces very many aspects of quantum theory, shows that these assumptions are not necessarily satisfied in practice.

It is interesting that a specialization of my own theory is closely related the Spekkens toy model, and this in itself indicates that my theory is related to several aspects of quantum mechanics.

This book is the result of a long process. In Helland (2006) an approach towards quantum mechanics was made in a leading statistical journal. Part of this approach is implemented later in the present book. In Helland (2008) an approach was made in a good physical journal, but again I know now that the reasoning is not complete. In the book Helland (2010) it was attempted to have two cultures in the mind at the same time. The book contains some relatively deep results in group action theory and in mathematical statistics, but the attempts made there to prove a link to quantum theory are too simple. The main limitation of all these three references is that I there attempted to deduce quantum theory from a version of statistical theory. In the present book I just assume that the two theories can be seen to have a common basis.

1.2 Different Views on the Foundation of Quantum Mechanics

The ordinary textbook formulation of quantum mechanics is very abstract. Its starting point: 'The state of a physical system is a normalized vector in a separable Hilbert space' has lead to an extremely rich theory, a theory which has not been refuted by any experiment and whose predictions range over an extremely wide variety of situations. Nevertheless, it is still unclear how this state concept should be interpreted.

Many conferences on quantum foundation have been arranged in recent years, but this has only implied that the number of new interpretations have increased, and no one of the old have died out. In two of these conferences, a poll among the participants was carried out (Schlosshauer et al. 2013; Norsen and Nelson 2013). The result was an astonishing disagreement on several simple and fundamental questions. One of these questions was whether quantum theory should be interpreted as an objective theory of the world (the ontological interpretation) or if it only expresses our knowledge of the world (the epistemic interpretation). According to Webster's Unabridged Dictionary, the adjective 'epistemic' means 'of or pertaining to knowledge, or the conditions for acquiring it'.

Recently Fuchs (2010), Fuchs and Schack (2011), Fuchs et al. (2013) and others have argued for various versions of Quantum Bayesianism, a radical interpretation where the subjective observer plays an important part. See also the philosophical discussion by Timpson (2008), the popular account in von Baeyer (2013) and the recent book (von Baeyer 2016).

QBism is a way of thinking about science quite generally, not just quantum physics. To cite Mermin (2014):

'QBism maintains that my understanding of the world rests entirely on the experiences that the world has induced in me throughout the course of my life. Nothing beyond my personal experience underlies the picture that I have formed of my own external world.'

But on the other hand:

'Facile charges of solipsism miss the point. My experience of you leads to hypothesize that you are very much like myself, with your own private experience.'

In the communication between you and me—in general between human beings, scientists and others, we need commonly defined concepts. I will come back to this later in the book.

This book takes as a point of departure that in some sense or other the epistemic interpretation should be important for issues of the quantum world. The next question then arises: Can one find a new and more intuitive *foundation* of quantum theory, a foundation related to the epistemic interpretation? It is my view that such a foundation ought to have some relation to statistical inference theory, another scientific fundament, which gives tools for wide variety of empirical investigations, and which in its very essence is epistemic. It should also be a kind of decision theory, related to decisions taken in everyday life.

Since the classical Copenhagen interpretation, which can be made precise in slightly different ways, several groups of researchers have proposed different interpretations of quantum mechanics. An extreme view on this was recently given by Tammaro (2014), who claimed that all these interpretations were deficient. More precisely, Tammaro stated that no current interpretation was consistent with experiment, resolved the measurement problem, and was completely free from logical deficiencies or fine-tuning problems.

Central to this discussion was the measurement problem, in particular the reconciliation of the two possible modes of change that the wave function can take: (1) Discontinuous, indeterministic time evolution sending $|\psi\rangle$ into an eigenstate $|o_i\rangle$ of observable O as a result of measurement of O. (2) Unitary time evolution governed by the Schrödinger equation.

Tammaro claimed to demonstrate that these two processes are inconsistent. His argument considered the state of the observer plus system after a measurement by the observer has been made. Process (1) then generates a mixed state, while Process (2) generates a pure state. One possible view is as follows: That this represents an inconsistency, may be related to the assumption that this quantum state really represents some objective reality for the system plus observer. According to QBism, a quantum state is always connected to some agent, and at each time it represents the subjective reality of that agent. In particular, the above system plus observer may be observed by another agent, and the wave function of this 'Wigner' (relative to the first agent, 'Wigner's friend') at time t may represent the belief or knowledge that he has at time t. This may depend upon how he obtains knowledge about the system, the agent, and the system plus agent.

Discussions in the literature of the EPR paradox (Einstein et al. 1935) and of Bell's theorem (Bell 1987), usually end up with the statement that quantum mechanics must violate the assumption of local realism. Since the locality assumption is inherent in relativity theory, my view is that it is the assumption of realism which must be discussed further. The quantum state represents the subjective reality of an agent or by a group of communicating agents, and this is in principle all there is to it. When all real and imagined agents agree on some observation at time t, this may be regarded as an objective reality at time t.

This is a rather radical view upon what reality should mean to us, but it is not inconsistent with the fact that people in complex macroscopic situations also may have different world views, and that these different world views may be extremely difficult to reconcile. This analogy should not be taken too far, though. The agents of quantum theory are ideal observers, while the humans of the macro world are far from perfect. The QBism interpretation of quantum mechanics is still only held by a minority of physicists.

My own views on quantum mechanics follow the views of the QBists a long way, but I differ in two respects: (1) For the purpose of a given experiment, the single observer can be replaced by a group of communicating observers. Thus language and the forming of concepts are important issues. (2) As formulated originally, QBism was closely tied to the philosophy of Bayesianism. I want to look upon the inference from observation in physics as related to inference in statistics, and I want to allow also other philosophies behind statistical inference. The ideal observer might well be a Bayesian, but since we humans are imperfect, we must also be allowed to use frequentist methods.

In any case, both the QBists and I agree that epistemic processes are of importance in the understanding of quantum mechanics. The simplest epistemic process consists of at least two decisions by the relevant agent (or communicating agents): First a decision on what question to nature to focus upon. Then the experiment or collection of data itself. Then finally a decision on how to deduce from these data the answer to the question originally posed.

Statistical theory and practice are almost solely concentrated on the last decision here. The decisions connected to focusing are attempted included in the present book.

An interesting discussion of various interpretations of quantum mechanics can be found in Khrennikov (2014). Khrennikov (2016) discusses QBism from the point of view of general decision making.

1.3 Theory of Decisions: Focusing—Context

Classical decision theory has been used with great success in a variety of fields like economics, medicine and politics. It is the basis for much of statistical inference theory. Yukalov and Sornette (2008, 2010, 2011, 2014) have in a series of papers tried to challenge this tradition with their Quantum Decision Theory (QDT). QDT is based upon the formalism of separable Hilbert spaces. It is parallel to quantum theory in many respects, but this does not imply that the decision maker is a quantum object. QDT is a way to avoid dealing with hidden variables, but at the same time reflecting the complexity of nature. The authors demonstrate that several paradoxes of classical decision theory can be resolved within QDT, and they claim that QDT covers both conscious and unconscious decisions.

This is not a place to describe QDT in detail, but the main idea is a mindspace $\mathcal{M}$ spanned by states corresponding to elementary prospects e_n. These elementary prospects are intersections of intended actions. Both prospect states $|\pi_j\rangle$, describing possible future actions, and strategic states $|\psi_s(t)\rangle$, describing the actor at time t, are vectors in $\mathcal{M}$. Prospect probabilities are given as $p(\pi_j) = |\langle \pi_j | \psi_s(t) \rangle|^2$, and rational decision makers maximize these probabilities.

QDT is presented as a formalism by Yukalov and Sornette, but there is also some discussion in the articles cited above of possible intuitive reasons behind this formalism. It is interesting in this connection that there is a large recent literature on various quantum models in psychology and cognitive science in books and articles; see Khrennikov (2010), Busemeyer and Bruza (2012), Bagarello (2013), Haven and Khrennikov (2013), Yukalov and Sornette (2009), Sornette (2014), Ashtiani and Azgomi (2015), Haven and Khennikov (2016) and in particular the review and discussion article (Pothos and Busemeyer 2013).

Other approaches to decision making using quantum theory methods are Aerts et al. (2014) and Eichberger and Pirner (2017).

It is clear that decisions can be made by single actors or by groups of communicating actors. Therefore language and a common set of concepts must be of some importance in a theory of decisions. In some cases decision making may take time; in other cases one does not have so long time in making decisions. Decisions may take into account selected experiences with past events, and may have a view towards future events. All decisions are made in a context. This context may be physical, historical, conceptual or constituting properties of the decision makers themself. In particular all these considerations are relevant for decisions made under an epistemic process.

1.4   The PBR Theorem. A Toy Model

Recently, Pusey et al. (2012) proved that the wave function must be ontic (i.e. a state of reality) in a broad class of realistic approaches to quantum theory. Two assumptions are made in that paper: (1) A system has a real physical state, not necessarily completely described by quantum theory, but objective and independent of the observer. (2) Systems that are prepared independently have independent physical states.

The assumption (1) goes to the roots of the traditional physicist's world view. I will claim that this state concept is unclear to many people outside the physical community. Why should one always be able to talk about a state independent of the observer? Of course the world itself exists independent of any observer, but the state of the world, what is that? As is discussed in Chap. 6 below, different people experience the world differently, also in macroscopic cases.

The Pusey, Barrett and Rudolph (PBR) theorem, related theorems and arguments connected to the theorem have been thoroughly reviewed by Leifer (2014). In particular it is discussed in detail there what is meant by a realistic approach. It is admitted that there are views of quantum mechanics that are not realistic, and that the Pusey, Barrett, Rudolph theorem does not apply to such interpretations. The approach discussed in the present book belongs to this class, broadly characterized in Leifer (2014) as neo-Copenhagen views. A recent argument against a realistic interpretation of the wave function is given by Rovelli (2016). A new criticism of the PBR theorem, discussing the ontological models framework, is given in Charrakh (2017).

The different interpretations of quantum theory were recently attempted classified by Cabello (2015). First, the interpretations were divided into two types: Type I (intrinsic realism) and Type II (participatory realism). For further discussion of the concept of participatory realism, see Fuchs (2016). The Type II interpretations were further divided into those concerned about knowledge and the one concerned about belief (QBism). As will become clear, this book is concentrating on Type II interpretations concerned about knowledge. According to Cabello (2015), this includes among others the classical Copenhagen interpretation, the approach by Zeilinger (1999) and even the no 'interpretation' approach by Fuchs and Peres (2000).

As a possible motivation behind epistemic views of quantum mechanics, the toy model of Spekkens (2007) is based on a principle that restricts the amount of knowledge an observer can have about reality. A wide variety of quantum phenomena were found to have analogues within this toy theory, and this can be taken as an argument in favour of the epistemic view of quantum states.

In the simplest version of the toy model, we have one elementary system. This system can be in one of the four ontic states 1, 2, 3 or 4, but our knowledge of this is in principle restricted. We can only know one of the following six epistemic states: (a) The ontic state is 1 or 2; (b) it is 3 or 4; (c) it is 1 or 3; (d) it is 2 or 4; (e) it is 1 or 4; or (f) it is 2 or 3. These are the epistemic states of maximal knowledge.

The ontic base of the state a) is {1, 2} etc. If the intersection of the ontic bases of a pair of epistemic states is empty, then those states are said to be disjoint. Thus (a) and (b) are disjoint, (c) and (d) are disjoint, and (e) and (f) are disjoint. There is a correspondence with certain basis vectors of the two-dimensional complex Hilbert space, where disjointness corresponds to orthogonality in the Hilbert space. For those who knows the Bloch sphere representation of that Hilbert space, the pairs of disjoint epistemic states can be pictured on the intersections of three orthogonal axes with that sphere.

Transformations of the epistemic states correspond to permutations of the ontic states. Thus the underlying group is the permutation group of four symbols, which has 24 elements. Each permutation induces a map between the epistemic states. In the Hilbert space correspondence, the even permutations correspond to unitary transformations, and the odd permutations correspond to anti-unitary transformations.

The toy model of Spekkens (2007) is generalised in several directions in Spekkens (2014). The generalisations are called epirestricted theories, and are showed to be equivalent to subtheories of quantum mechanics. An epirestricted theory consists of three steps. One starts with a classical ontological theory. Then one constructs a statistical theory over these ontic states. Finally one postulates a restriction on what sorts of statistical distributions that can describe an agent's knowledge of the system. The theory of the present paper is related both to this and to the QBism school, but there are important differences, as will be seen from the discussion below.

1.5 Epistemic Processes

The Quantum Bayesianism is founded on an observer's belief, quantified by a Bayesian probability. I want to relate my state concept also to the notion of *certain* belief, which I call knowledge. The knowledge will be associated with an agent or with a group of communicating agents, and his/her/their knowledge will be knowledge about what I will call an e-variable.

An epistemic process is any process under which an agent or a group of communicating agents obtain knowledge about a physical system. In general there are many ways by which one can obtain knowledge about the world or about aspects of the world. In a given situation the observer has some background, in terms of his history, in terms of his physical environment, and in terms of the concepts that he is able to use in analysing the situation. This is called the context of the observer, and the context may limit his ability to obtain knowledge.

A *conceptual variable* is any variable related to the physical system, defined by an agent or by a group of communicating agents. The variable may be a scalar, a vector or belong to a larger space.

An *e-variable* or epistemic conceptual variable θ is a conceptual variable associated with an epistemic process: Before the process the agent (or agents) has (have) no knowledge about θ; after the process she/they has/have some knowledge, in the simplest case full knowledge: $\theta = u_k$. Here u_k is one of the possible values that θ can take. In this book it is mostly assumed for simplicity that θ is discrete, which it will be in the elementary quantum setting below. For a continuous variable θ, knowledge on the e-variable will be taken to mean a statement to the effect that θ belongs to some given set, an interval if θ is a scalar.

The e-variable concept is a generalization of the parameter concept as used in statistical inference, introduced by Fisher (1922), and today incorporated in nearly all applications of statistics. In statistics, a parameter θ is usually an index in the statistical model for the observations, and the purpose of an empirical investigation is to obtain statements about θ, in terms of point estimation, confidence interval estimation or conclusion from the testing of hypotheses. The parameter is often associated with a hypothetical infinite population. My e-variable will also be allowed to be associated with a finite physical system, a particle or a set of particles. But, in the same way as with a parameter, the purpose of any empirical investigation will be to try to conclude with some statement, a statement expressed in terms of an e-variable θ.

1.5.1 E-Variables in Simple Epistemic Questions

The point of departure is that we ask a question to nature, a question in order to achieve increased knowledge. The e-variable is the particular conceptual variable that is connected to such a question.

To give a very simple example, let us assume that we are given some object A, and ask 'What is the weight of object A?'. Then $\mu =$ 'weight of A' is an e-variable. We can use a scale to obtain a very accurate estimate of μ. Or we can use several independent measurements, and use the mean of those as a more accurate estimate. In the latter case it is common to introduce a statistical model where μ is a parameter of that model. But in my view the e-variable concept is a more fundamental notion. The variable μ exists before any statistical model is introduced. Most people will agree that μ exists in some sense. Thus in this example the e-variable has some ontic basis, but my claim is that this need not always be the case in all epistemic processes. Even in this case the existence of μ as a real number may be discussed. For instance, the question 'Is μ rational or irrational?' is rather meaningless.

In some cases we can obtain very precise knowledge about some e-variable after some time. As an example, ask 'What will be the number of sun hours tomorrow?' To to answer this question today, we need huge computer models and expert advice from meteorologists. Tomorrow, it is just a question about counting the sun hours.

Similar situations occur when doing simple causal inference. We may for instance ask 'What is the causal effect of medicine C on individual A?'. The e-variable we might have in mind, might be 'Time to recovery for A' compared to time to recovery without any medicine. If we have no earlier experience with A having the disease in question, we have a counterfactual situation. If we have observed A earlier with the same disease when he did not take any medicine, we can at least have a tentative answer within a few days.

1.5.2   E-Variables in Statistics

The most important kinds of e-variables in statistics are statistical parameters. The concept of a statistical parameter was introduced by Fisher (1922), and is today used in nearly all cases where statistical inference is applied. A statistical model is a probability model of the data given the parameters, and the purpose of inference is to obtain information about the parameters. Sometimes only a subset of the parameters are of interest, and the epistemic question is then 'What are the values of these parameters?' Usually one will not be able to get complete information. Partial information, given the data, can be expressed in terms of confidence regions of credibility regions, concepts that will be discussed later in this book.

A completely different kind of e-variables—here called simple e-variables—occur in prediction problems. When we want to predict a random variable Y, the epistemic question is again 'What is the value of Y?' The answer is often sought via a statistical model by first estimating the parameters of the model, and then formulating a prediction equation from this. This gives a predicted value $\widehat{Y}$, again incomplete information from the data. But prediction problems can also be seen as of different nature than estimation problems, requiring separate techniques.

1.5.3   E-Variables in Causal Inference

A causal model is different from a statistical model, as stressed by Pearl (2009). Statistical concepts are correlation, regression, conditional independence, association, likelihood etc., all concepts that can be related to a statistical model. Causal concepts are randomization, influence, effect, confounding etc.. A causal model is defined in terms of what is called a directed acyclic graph, and again these models contain parameters. The epistemic question is again 'What is the value of these parameters?' Sometimes, but not always, incomplete answers can be given by means of data.

Simple e-variable also occur in causal inference. Again we can look at the example 'What is the causal effect of medicine C on individual A?', and the e-variable might be 'Time to recovery for A'.

1.5.4 E-Variables in Quantum Mechanics

E-variables as statistical parameters connected to an infinite population may also occur in quantum mechanics. An example is given in Wootters (1980). Consider a photon which has just emerged from a polarizing filter and which is about to encounter a Nicol prism. The filter can have a continuum of possible orientations, each characterized by a polarization angle θ. But when the photon encounters the Nicol prism, it is required to choose between exactly two possible actions: (1) to go straight through the Nicol prism, or (2) to be deflected in a direction uniquely determined by the orientation of the prism. A straightforward epistemic problem is to estimate θ form an ensemble of photons using the probability law $p(\theta) = \cos^2\theta$. Wootters was investigating the much deeper problem whether this probability law is the best possible in some sense.

But in quantum mechanics simple e-variables are most important. A quantum system can be given some preparation, and under this preparation e-variables like position, momentum, energy, spin, angular momentum may be investigated. The quantum formulation is introduced by associating each e-variable to a Hermitian operator in a Hilbert space (in the discrete case to be precise; for continuous variables a more general construction is needed, one approach is the rigged Hilbert space, see Ballentine 1998) In the present book I will mostly concentrate on discrete e-variables, at least to begin with, and these e-variables always correspond to Hermitian operators in some basic Hilbert space. The quantum states are in general given by vectors in the basic Hilbert space, but in this book special emphasis is often given to state vectors that are eigenvectors of some Hermitian operator. Given a quantum system and an e-variable for this system, a natural focused question will be 'What is the value of this e-variable?', and a simple (ideal) measurement will give the answer. But the epistemic question can also be to predict the e-variable before the measurement is done. Then an incomplete answer can be given, the probability distribution as found from the Born rule.

Many books and papers use the term 'observable' for what I have called a simple e-variable in quantum mechanics. Ballentine (1998), remarking on some ambiguity in the use of this term, prefers to use 'dynamical variable'. Bell (1975) introduced the term 'beable' for a related concept, assuming some sort of reality of the dynamical variables. My own point is basically to relate the e-variable concept closely to an epistemic process.

The first basic postulate of the quantum formalism may now be written:

Postulate 1.1 *To each physical system there corresponds a Hilbert space. To each simple e-variable of this physical system there corresponds a unique Hermitian operator of the Hilbert space. The possible values of the e-variable are the eigenvalues of this operator.*

(This formulation is only valid for discrete e-variables, which will occupy most of this book. In general one needs a rigged Hilbert space if possible values as eigenvalues should be taken literally.)

There are several recent attempts to motivate this postulate from more intuitive assumptions, see references later. My own attempt, starting from what I call a symmetrical epistemic setting, generalizing the case of spin/ angular momentum, is contained in Chap. 4 below.

1.5.5 Real and Ideal Measurements in Quantum Mechanics

As emphasized by Ballentine (1998), it is important to distinguish between preparation and measurement of a physical system. In my notation from Chap. 3, the preparation gives part of the context τ for the measurement. The purpose of the measurement is to say something about an e-variable θ^a.

To be concrete, let us assume that we want to measure the z-component θ^z of the spin of a silver atom. (A similar experiment with a charged particle like electron will lead to practical difficulties due to the so-called Lorentz force acting upon the charge.) The spin component takes one of two possible values, say ± 1. The experiment is done by sending the atom in some direction, say the y-direction, through a magnetic field which is inhomogeneous in the z-direction. This will cause the atom to be deflected up in the z-direction if $\theta^z = +1$, down if $\theta^z = -1$.

How should this deflection be detected? One way would be to place a screen of detectors in the xz-plane after the silver atom has passed the magnetic field. Hopefully this will give a click in one detector in the positive z-direction if $\theta^z = +1$, or one in the negative z-direction if $\theta^z = -1$.

But such detectors are far from perfect. it may be that the silver atom goes through the screen without being detected. It may be that it clicks on two or more neighbouring detectors when passing through the screen. What we have as a result of the experiment is *data*, which is an array of 0's and 1's corresponding to the detectors: 1 if the detector clicks, 0 otherwise.

This kind of detection errors may seem like a nuisance, but they have fundamental physical importance. Similar detection errors have played a large role in the physical literature on Bell's inequality, which will be discussed in Sect. 5.8 below. The point here is that one on the background of such detection errors have proposed possible 'loopholes' in quantum mechanical experiments to test Bell's theorem. Very recent experiments have excluded such loopholes.

Another point about this simple experiment is the following: The same magnet and screen of detectors can be rotated so as to measure spin component θ^a in any direction a, and the same probability model can be used for the measurement. Thus there is strictly speaking no reason to use a superscript a on the data z^a of the experiment, but we will do it anyway to make it clear that this is the data for the experiment for the spin e-variable θ^a in the direction a.

Thus we have data z^a, and we may construct a statistical model for these data, a model depending on the true value of θ^a. The details of this model need not concern us now, but my important point is that we are in the generalized experiment situation discussed in detail in Sect. 3.2. The distribution of z^a, given the context τ, depends on the unknown e-variable θ^a, and relative to this distribution, τ is independent of θ^a. (A prior for θ^a may depend upon τ, however.)

So far, we have had only one particle, but a similar construction applies in the situation with a beam of n independent particles. Then we have a set of e-variables $\theta^a = (\theta_1^a, \theta_2^a, \ldots, \theta_n^a)$ with data $z^a = (z_1^a, z_2^a, \ldots, z_n^a)$. When $n \to \infty$, this approaches a probability distribution over the e-variable, and the data may without loss of generality be reduced to a frequency distribution of clicks over the detectors of the screen. Under suitable assumptions the probability distribution of the e-variable is given by the Born rule derived in Chap. 5, and the data reduction corresponds to a reduction by sufficiency, a concept discussed in Chap. 3.

In Chap. 3 three fundamental principles of statistical inference are discussed, the conditionality principle, the sufficiency principle and the likelihood principle, where the last one is shown to follow from the first two principles. This whole apparatus is now available for experiments of the quantum type. I will in fact use the likelihood principle later when deriving Born's rule, the basic probability rule of quantum theory. It is therefore important to me that a similar discussion with an imperfect apparatus in principle can be carried out in all quantum experiments. But this is in principle. Once the statistical principles are established, we can return to ideal experiments were data are in one-to-one correspondence with the e-variables of the experiment.

The same kind of data and data model can be used for any choice of focused direction, focused measurement. Assume that the situation is such that the vector of different e-variables θ^a is inaccessible; see Sect. 1.5.8. Then this gives rise to a focused version of the likelihood principle. This is the kind of likelihood principle used to derive the Born rule. The details are in Chaps. 3, 4 and 5.

1.5.6   Quantum States, Their Interpretations, and a Link to the Ensemble Interpretation

Pure quantum states are defined formally as unit vectors in the Hilbert space. These vectors may or may not be eigenvectors of physical meaningful Hermitian operators. They are denoted as ket vectors $|k\rangle$ and may be assumed to form a complete set. For more information, see Sect. 5.1 and Appendix B.

Let first $|k\rangle$ be a unit eigenvector of a Hermitian operator A corresponding to a simple e-variable θ. Assume for simplicity that A do not have multiple eigenvalues. Then, from an epistemic process point of view, and from what can be seen as a simple observation from the quantum formalism, $|k\rangle$ may be identified by a question 'What is the value of θ?' together with a definite answer '$\theta = u_k$', where u_k is the corresponding eigenvalue.

Given a unit vector $|k\rangle$, there will in general be an infinity of Hermitian operators for which $|k\rangle$ is an eigenvector, and there may be many such operators which correspond to physically meaningful e-variables θ. Also, one must consider the case with multiple eigenvalues. Hence, given $|k\rangle$, its association to a question-and-answer pair is not unique in general. In the special case of the spin or angular momentum of a particle and in related cases, there is some uniqueness, however.

(a) For a spin 1/2 particle, or more generally for a qubit, a unit vector in a 2-dimensional Hilbert space, there is a unique question-and-answer pair for spin component corresponding to each unit vector; see Sect. 5.1.1 and Proposition 5.4 of Sect. 5.2.
(b) In the general case of spin or angular momentum component there will correspond a unique normalized ket vector corresponding to each question-and-answer pair; see Proposition 5.3 of Sect. 5.2. For dimension higher than 2 there are however unit vectors that do not correspond directly to a question-and-answer pair for spin/angular momentum components.
(c) In a general setting, related to, but weaker than that of spin/ angular momentum, one can prove without assuming a quantum formalism from the outset, that each question-and-answer pair under a certain technical assumption corresponds to a unique normalized ket vector of some Hilbert space. This is proved in Chap. 4. Unfortunately, one of the technical assumptions as stated there, seems to be too strong; it is argued that it should be possible to weaken this assumption.

A larger class of pure states than those corresponding directly to question-and-answer pairs for some given set of e-variables may be found by taking linear combinations, which may be motivated physically, or as solutions of the Schrödinger equation.

In most of this book I will discuss epistemic processes, processes to obtain knowledge about some system, physical or otherwise. Then the question-and-answer pair corresponds to two decisions: First a decision on which e-variable to focus upon. Then after the data are obtained, a decision on the value of this e-variable. In statistical theory, only the last of these decisions is discussed. It is very interesting that the quantum state concept turns out to be useful for more general decisions; from a qualitative point of view this will be discussed in Chap. 6 below.

To the ket vector $|k\rangle$ correspond the bra vector $\langle k|$ and the projection operator $|k\rangle\langle k|$.

The more general concept of a mixed state is defined from a probability distribution $\{\pi_k\}$ as follows:

$$\rho = \sum_k \pi_k |k\rangle\langle k|. \qquad (1.1)$$

Depending upon the observator(s) and upon the physical situation, the probabilities π_k may be interpreted—and assessed/estimated—in three possible ways: (1) as Bayesian prior distributions; (2) as Bayesian posterior distributions; (3) as frequentist confidence distributions, see Schweder and Hjort (2016). From a statistical point of view, Bayesian probabilities are connected to credibility intervals, confidence distributions are derived from confidence intervals; see Chap. 2 below.

It is interesting that under specific symmetry assumptions, the confidence intervals and the credibility intervals coincide and are associated with the same probability; see Corollary 3.6.2 p. 93 in Helland (2010). A consequence of this is that the probabilities estimated under the interpretations (2) and (3) will be numerically equivalent under these symmetry assumptions.

In this section it will be convenient to use the confidence distribution interpretation (3), even though it is new and unknown, also largely among statisticians. The basic concept is that of a confidence interval with some variable confidence coefficient; details will be discussed in Chap. 2. The important point to us now is the interpretation of the confidence coefficient. It can be seen as a limiting frequency obtained from data in a large hypothetical set of epistemic processes were the same statistical method is used.

A similar idea carries over to the quantum situation: Ballentine (1998) makes an important point of the fact that probabilities in quantum mechanics must be interpreted with respect to a hypothetical ensemble. In Smolin (2011) it is proposed that this hypothetical ensemble is realised by all systems in the universe that occupy the same quantum state.

The state concept is the basis for all calculation of probabilities in quantum mechanics. A physical system is prepared in some given state, and probability distributions of measurements from this state are given by the Born rule.

The Born rule is developed from my point of view in Sect. 5.6 below. A consequence of this rule is a simple formula which can be used to restate Ballentine's

Postulate 1.2 *To each state there corresponds a unique state operator ρ. The average value of an e-variable θ, represented by the operator A, in the virtual ensemble of events that may result from a preparation procedure for state represented by the operator ρ is*

$$\langle \theta \rangle = \mathrm{tr}(\rho A).$$

From a statistical point of view, the average value here is interpreted as an expectation. By extending this formula in a natural way, the probability distribution of θ under ρ may be derived; see again Ballentine (1998). In fact this probability distribution follows easily directly from my Theorem 5.6 in Sect. 5.6.

1.5.7 Quantum States for Spin 1/2 Particles

Consider again the silver atom of Sect. 1.5.5, and suppose that we want to measure the spin component in some direction a. Let us now assume that the measurement apparatus is perfect. Then the experiment has two possible outcomes: spin up or spin down. This corresponds to two possible quantum states:

$$|a; +\rangle \quad \text{or} \quad |a; -\rangle. \tag{1.2}$$

Let now a be arbitrary. In Sect. 5.2 I will prove the important result: *All pure quantum states for spin 1/2 particles can be written in one of the forms (1.2) for some a.*

In concrete terms, this means that every pure quantum state for a spin 1/2 particle can be interpreted as a question: What is the spin component in direction a? together with a definite answer: $+1$ or -1. (Strictly speaking, this gives a double counting of quantum states, since, given some direction a, we can also choose the opposite direction $-a$; and $\theta^a = +1$, say, is equivalent to $\theta^{-a} = -\theta^a = -1$.)

It is of some related interest that the axioms of quantum mechanics recently were approached by Smilga (2017) by taking spin measurements in different directions as a point of departure, and using group representation theory.

1.5.8 Inaccessible Conceptual Variables and Complementarity

It is important that not all conceptual variables are e-variables. A conceptual variable ϕ is called *inaccessible* if there is no epistemic process by which one can get accurate knowledge about it. An example from the area of quantum mechanics is $\phi = (\xi, \pi)$, where ξ is the position of a particle, and π is the momentum. An example which will be discussed throughout the present book, is $\phi = (\lambda^x, \lambda^y, \lambda^z)$, where λ^a is the component of an angular momentum or a spin for a particle or a system of particles, the component in direction a. Here each λ^a is an accessible conceptual variable, an e-variable, but the vector ϕ is inaccessible. Also macroscopic examples abound, for instance connected to counterfactual situations in causal inference: Let θ^1 be the time to recovery for given patient A at time t when he is given treatment B, let θ^2 be his time to recovery when he is given treatment C, and let $\phi = (\theta^1, \theta^2)$.

In such cases, where a vector of e-variables is inaccessible, it is equivalent to say that the components are complementary. Since introduced by Bohr, the concept of complementarity has played a fundamental and important role in quantum mechanics.

It is essential to stress that the conceptual variables above are not hidden variables. Variables like ϕ are just mathematical variables, but variables upon which group actions may be defined. $\phi = (\xi, \pi)$ may be subject to Galilean transformations, time translations or changes of units, while $\phi = (\lambda^x, \lambda^y, \lambda^z)$ may be subject to rotations. This will of course also induce transformation of the components, the e-variables. Important transformations in the group of rotations of the last ϕ are: (1) Those leading to a change in the values of λ^x (or of any other fixed component); (2) Those leading to an exchange of λ^x and λ^y (or any other pair of components).

The (simple) e-variables are not hidden variables, but closely tied to the epistemic processes.

To illustrate the general view of this book, look at Fig. 1.1. Here ϕ is an inaccessible conceptual variable, the θ's are e-variables, the τ's are context variables

Fig. 1.1 A graphical picture,
illustrating a general view
upon quantum theory

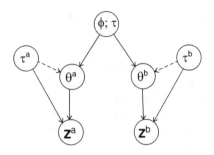

and the z's are data. The upper arrows denote functional dependence, and the
lower arrows denote conditional probability distributions of the data. The dotted
arrows indicate that one may or may not have a prior for θ, given the context.
The experimentalist has the choice between two mutually excluding experiments,
denoted by a and b.

When does this situation lead to a quantum theory, which can alternatively be
described by a Hilbert space formulation? A partial answer is given in Chap. 4 and
in Sect. 5.2 below. The crucial concept is the context variable τ. One possibility
is that this denotes the maximal symmetrical epistemic setting of Chap. 4 (here
θ is replaced by λ), satisfying Assumptions 4.1–4.3 there. Another possibility is
the corresponding general symmetrical epistemic setting, and a final situation is
a spin/angular momentum situation. An open question is to find the most general
conditions under which a quantum theory can come into being from the situation of
Fig. 1.1. In the situations above, the θ's are discrete, which they are in elementary
quantum theory, but we can also let the same figure illustrate a setting where the θ's
are continuous; for a completely general formulation, see Sect. 5.15.

The inference on simple e-variables as presented in this book implies some
connection between the quantum mechanical culture and the statistical culture. In
the next chapter I will give a broad description of the science of statistical inference.
Then I will give some general principles on inference on parameters/ e-variables
when data are involved, and after that I will turn to quantum theory.

References

Aerts, D., Sozzo, S., & Tapia, J. (2014). Identifying quantum structures in the Ellsberg paradox.
   International Journal of Theoretical Physics, 53, 3666–3682.
Ashtiani, M. B., & Azgomi, M. A. (2015). A survey of quantum-like approaches to decision
   making and cognition. *Mathematical Social Sciences, 75*, 49–80.
Bagarello, F. (2013). *Quantum dynamics for classical systems*. Hobroken, NJ: Wiley.
Ballentine, L. E. (1998). *Quantum mechanics. A modern development*. Singapore: World Scientific.
Bell, J. S. (1975). The theory of local beables. Reprinted in Bell (1987).
Bell, J. S. (1987). *Speakable and unspeakable in quantum mechanics*. Cambridge: Cambridge
   University Press.
Brody, T. (1993). In L. de la Pera & P. Hodgson. *The philosophy behind physics*. Berlin: Springer.

Busemeyer, J. R., & Bruza, P. (2012). *Quantum models of cognition and decision*. Cambridge: Cambridge University Press.

Cabello, A. (2015). Interpretations of quantum theory: A map of madness. arXiv: 1509.0471v1 [quant-ph].

Charrakh, O. (2017). On the reality of the wavefunction. arXiv: 1706.01819 [physics.hist-ph].

Einstein, A., Podolsky, B., & Rosen, N. (1935). Can quantum-mechanical description of physical reality be considered complete? *Physical Review, 47*, 777–780.

Eichberger, J., & Pirner, H. J. (2017). Decision theory with a Hilbert space as a probability space. arXiv: 1707.07556 [quant-ph].

Fisher, R. A. (1922). On the mathematical foundations of theoretical statistics. *Philosophical Transactions of the Royal Society of London. Series A, 222*, 309–368. Reprinted in: Fisher R. A. Contribution to Mathematical Statistics. Wiley, New York (1950)

Fuchs, C. A. (2010). QBism, the Perimeter of Quantum Bayesianism. arXiv: 1003.5209v1 [quant-ph].

Fuchs, C. A. (2016). On participatory realism. arXiv: 1601.04360v2 [quant-ph].

Fuchs, C. A., Mermin, N. D., & Schack, R. (2013). An introduction to QBism with an application to the locality of quantum mechanics. arXiv: 1311.5253v1 [quant-ph].

Fuchs, C. A., & Peres, A. (2000). Quantum theory needs no interpretation. *Physics Today, S-0031-9228-0003-230-0*; Discussion *Physics Today, S-0031-9228-0009-220-6*.

Fuchs, C. A., & Schack, R. (2011). A quantum-Bayesian route to quantum-state space. *Foundations of Physics, 41*, 345–356.

Haven, E., & Khrennikov, A. (2013). *Quantum social science*. Cambridge: Cambridge University Press.

Haven, E., & Khennikov, A. (2016). Quantum probability and mathematical modelling of decision making. *Philosophical Transactions of the Royal Society A, 374*, 20150105.

Helland, I. S. (2006). Extended statistical modeling under symmetry; the link toward quantum mechanics. *Annals of Statistics, 34*, 42–77.

Helland, I. S. (2008). Quantum mechanics from focusing and symmetry. *Foundations of Physics, 38*, 818–842.

Helland, I. S. (2010). *Steps towards a unified basis for scientific models and methods*. Singapore: World Scientific.

Khrennikov, A. (2010). *Ubiquitous quantum structure*. Berlin: Springer.

Khrennikov, A. (2014). *Beyond quantum*. Danvers, MA: Pan Stanford Publishing.

Khrennikov, A. (2016). Quantum Bayesianism as a basis of general theory of decision making. *Philosophical Transactions of the Royal Society A, 374*, 20150245.

Knorr Cetina, K. (1999). *Epistemic cultures. How the sciences make knowledge*. Cambridge, MA: Harvard University Press.

Leifer, M. S. (2014). Is the quantum state real? An extended review of ψ-ontology theorems. arXiv:1409.1570v2 [quant-ph].

Mermin, N. D. (2014). Why QBism is not the Copenhagen interpretation and what John Bell might have thought of it. arXiv.1409.2454 [quant-ph].

Norsen, T., & Nelson, S. (2013). Yet another snapshot of fundamental attitudes toward quantum mechanic. arXiv:1306.4646v2 [quant-ph].

Pearl, J. (2009). *Causality. Models, reasoning and inference* (2nd ed.). Cambridge: Cambridge University Press.

Pothos, E. M., & Busemeyer, J. R. (2013). Can quantum probability provide a new direction for cognitive modeling? With discussion. *Behavioral and Brain Sciences, 36*, 255–327.

Pusey, M. F., Barrett, J., & Rudolph, T. (2012). On the reality of quantum states. *Nature Physics, 8*, 475–478.

Rovelli, C. (2016). An argument against a realistic interpretation of the wave function. *Foundations of Physics, 46*, 1229–1237.

Schlosshauer, M., Koer, J., & Zeilinger, A. (2013). A snapshot of fundamental attitudes toward quantum mechanics. *Studies in History and Philosophy of Modern Physics, 44*, 222–238.

Schweder, T., & Hjort, N. L. (2016). *Confidence, likelihood, probability. Statistical inference with confidence distributions.* Cambridge: Cambridge University Press.

Smilga, W. (2017). Towards a constructive foundation of quantum mechanics. *Foundations of Physics, 47,* 149–159.

Smolin, L. (2011). A real ensemble interpretation of quantum mechanics. aXiv. 1104.2822 [quant-ph].

Sornette, D. (2014). Physics and financial economics (1776-2014): puzzles, Ising and agent-based models. *Reports on Progress in Physics, 77,* 062001.

Spekkens, R. W. (2007). In defense of the epistemic view of quantum states: A toy theory. *Physical Review A, 75,* 032110.

Spekkens R. W. (2014). Quasi-quantization: Classical statistical theories with an epistemic restriction. arXiv.1409.304 [quant-ph].

Tammaro E. (2014). Why current interpretations of quantum mechanics are deficient. arXiv1408.2083v2 [quant-ph].

Timpson, C. G. (2008). Quantum Bayesianism: A study. *Studies in History an Philosophy of Modern Physics, 39,* 579–609.

von Baeyer, H. C. (2013). Quantum weirdness? It's all in your mind. *Scientific American, 308*(6), 38–43.

von Baeyer, H. C. (2016). *QBism: The future of quantum physics.* Harvard: Harvard University Press.

von Neumann, J. (1932). *Mathematische Grundlagen der Quantenmechanik.* Berlin: Springer.

Wootters, W. K. (1980). *The Acquisition of Information from Quantum Measurements.* PhD Thesis. Center for Theoretical Physics. The University of Texas at Austin.

Yukalov V. I., & Sornette, D. (2008). Quantum decision theory as a quantum theory of measurement. *Physics Letters A, 372,* 6867–6871.

Yukalov, V. I., & Sornette, D. (2009). Processing information in quantum decision theory. *Entropy, 11,* 1073–1120.

Yukalov, V. I., & Sornette, D. (2010). Mathematical structure of quantum decision theory. *Advances in Complex Systems, 13,* 659–698.

Yukalov, V. I., & Sornette, D. (2011). Decision theory with prospect interference and entanglement. *Theory and Decision, 70,* 383–328.

Yukalov, V. I., & Sornette, D. (2014). How brains make decisions. *Springer Proceedings in Physics, 150,* 37–53.

Zeilinger, A. (1999). A foundational principle for quantum mechanics. *Foundations of Physics, 29,* 631–643.

Chapter 2
Statistical Inference

Abstract A summary of statistical inference theory as it has been developed up to now, is given. Both frequentist and Bayesian inference are covered. The recent concept of confidence distributions is described. As a preparation for the next chapter, statistical inference is also seen as a special case of inference in a general epistemic process.

2.1 Basic Statistics

2.1.1 Probability

The concept of probability is basic for statistical theory. As developed by Kolmogorov in 1930, it is a taken as a normed measure on some probability space Ω.

Formally, we first introduce a σ-algebra $\mathcal{F}$ of subsets of Ω. The mathematical requirements are: (a) Ω should be in $\mathcal{F}$; (b) the complement A^c should be in $\mathcal{F}$ whenever A is in $\mathcal{F}$, where $A^c = \{\omega \in \Omega : \omega \notin A\}$; (c) $\cup_{n=1}^{\infty} A_n$ should be in $\mathcal{F}$ whenever A_n is in $\mathcal{F}$ for $n = 1, 2, \ldots$.

A normed measure P is then a set function such that a) $0 \leq P(A) \leq P(\Omega) = 1$ for all $A \in \mathcal{F}$; b) $P(\cup_{n=1}^{\infty} A_n) = \sum_{n=1}^{\infty} P(A_n)$ if the sets A_i and A_j are disjoint; i.e., $A_i \cap A_j = \emptyset$ when $i \neq j$. This implies $P(A^c) = 1 - P(A)$ and $P(A \cup B) = P(A) + P(B) - P(A \cap B)$ for all sets A and B. The sets $A \in \mathcal{F}$ are called events, and the triple $(\Omega, \mathcal{F}, P)$ is called a probability space.

If Ω is a topological space, the Borel σ-algebra is the smallest σ-algebra containing all open sets. If Ω is discrete and finite, we can, and will, take $\mathcal{F}$ to consist of all subsets.

A random variable X is a measurable function from Ω into the Euclidean space $\mathcal{R}^n$, that is, a function such that $\{X \in B\} = \{\omega \in \Omega : X(\omega) \in B\} = X^{-1}(B)$ is in $\mathcal{F}$ whenever B is a Borel set in $\mathcal{R}^n$. The probability distribution of X is defined by $P(X \in B) = P(X^{-1}(B))$.

Readers not willing to go into all these mathematical details may think of a random variable X as some variable with a distribution associated with it. In this book I will work with real-valued random variables of two kinds:

© Springer-Verlag GmbH Germany, part of Springer Nature 2018
I. S. Helland, *Epistemic Processes*, https://doi.org/10.1007/978-3-319-95068-6_2

- Discrete finite-valued random variables X with point probabilities $p(i) = P(X = i)$; $i = 1, \ldots, r$ satisfying $\sum_{i=1}^{r} p(i) = 1$.
- Continuous random variables X with $P(a \leq X \leq b) = \int_a^b f(x)dx$ for some probability density $f(x)$ satisfying $\int_{-\infty}^{\infty} f(x)dx = 1$.

From this we define expectation

$$\mu = \mathrm{E}(X) = \sum_{i=1}^{r} i p(i); \quad \mu = \mathrm{E}(X) = \int_{-\infty}^{\infty} x f(x)dx$$

$$\mathrm{E}(g(X)) = \sum_{i=1}^{r} g(i) p(i); \quad \mathrm{E}(g(X)) = \int_{-\infty}^{\infty} g(x) f(x)dx$$

and variance

$$\sigma^2 = \mathrm{Var}(X) = \mathrm{E}[(X - \mu)^2].$$

Two random variables X and Y are independent if $P((X \in A) \cap (Y \in B)) = P(X \in A)P(Y \in B)$ for all Borel sets A and B, with a natural generalization to several random variables. For discrete random variables, this is equivalent to $p_{X,Y}(i, j) = p_X(i)p_Y(j)$; for continuous random variables it is equivalent to $f_{X,Y}(x, y) = f_X(x)f_Y(y)$ with an obvious definition of the joint density.

The interpretation of the probability concept is important for applications. In the literature, three different, but related interpretations are given:

1. The principle of equally likely outcome: If there are r possible outcomes, each is given the probability $1/r$. This can immediately be applied to discrete finite-valued variables, and has examples in the tossing of a die, the tossing of a coin, in card games, in opinion polls etc. Below I will generalize the principle to random variables with a compact range, using the group concept. When the range is not compact, we need un-normed measures. This causes conceptual difficulties that I will not go too deeply into in the present book. I will return to the problem at some points, however.
2. The principle of odds making or subjective probability: The probability of an event A is found on the basis of how much a person is willing to pay for each outcome in a wager with the two outcomes A and A^c. This was introduced by de Finetti and Savage, and used by them as a foundation for Bayesian statistics.
3. The principle of long run frequency: If an experiment is repeated n times, the relative frequency of the event A is the number of times A happens, divided upon n. The probability of A is interpreted as the limit in some sense (see below) of the relative frequency as $n \to \infty$. I will indicate below that this interpretation always can be applied, and made precise, in situations where an experiment can be repeated an arbitrary number of times and the probability can be defined from other considerations.

In many concrete applications, not only one, but two or three of these interpretations may be relevant.

The concept of conditional probability can be given a precise mathematical definition using the notion of a Radon-Nikodym derivative. Specifically, if $\mathcal{B}$ is a sub-σ-algebra of $\mathcal{F}$, then we define $P(A|\mathcal{B})$ as the unique (up to a P-measure 0) $\mathcal{B}$-measurable function such that

$$\int_B P(A|\mathcal{B})P(d\omega) = \int_B I_A(\omega)P(d\omega) \qquad (2.1)$$

for all $B \in \mathcal{B}$. If $\mathcal{B}$ is generated by the disjoint events $\{B_i\}$, this is consistent with the definition that $P(A|B_i) = P(A \cap B_i)/P(B_i)$ whenever $P(B_i) \neq 0$.

Finally, asymptotic considerations may simplify statistics in cases where there are many observations. I will introduce three limit concepts in probability:

- Convergence in probability: $P(|Y_n - Y| > \epsilon) \to 0$ for all $\epsilon > 0$.
- Convergence almost surely: $P(\{\omega : Y_n(\omega) \to Y(\omega)\}) = 1$. This is stronger than convergence in probability.
- Convergence in law: $P(Y_n \leq y) \to P(Y \leq y)$ for all y where $F(y) = P(Y \leq y)$ is continuous. This is a property of the distribution functions rather than of the random variables, but it is related in several ways to the concept of convergence in probability. For instance, for convergence to a degenerate distribution, the two concept are equivalent, and when Y_n tends in law to Y and U_n in probability to c, then $Y_n + U_n$ tends in law to $Y + c$.

In statistical applications, we often have repeated observations, and thus a sample $X = (X_1, X_2, \ldots, X_n)$, where the X_i's are independent with the same distribution. Assuming these have finite expectation μ and finite variance σ^2, one can prove three limit laws for the mean $\bar{X}_n = n^{-1}\sum_{i=1}^n X_i$:

- The law of large numbers I: $\bar{X}_n$ converges in probability to μ as $n \to \infty$.
- The law of large numbers II: $\bar{X}_n$ converges almost surely to μ as $n \to \infty$.
- The central limit theorem: $\sqrt{n}(\bar{X}_n - \mu)$ converges in law to $N(0, \sigma^2)$, where $N(\xi, \sigma^2)$ in general is the continuous distribution with density

$$f(x) = \frac{1}{\sqrt{2\pi}\sigma}\exp(-\frac{(x - \xi)^2}{2\sigma^2}). \qquad (2.2)$$

For the first and the third law, see Lehmann (1999). The second law is proved for instance in Sen and Singer (1993).

Using the law of large numbers on the indicator functions $X_i = I(Z_i \in A)$, one easily shows that the frequency interpretation of the probability concept always is valid in situations where it is applicable.

2.1.2   Statistical Models

In general, a model is a representation of the real world, simplified, but designed such that the essential features that one is interested in, are focused in the model and are correctly represented in the model. A map of the London underground is sometimes taken as an example of a model.

In statistics, one wants a model which can be employed in the epistemic process. This is the reasoning used: The unknown feature that one is interested in, is modeled as a *parameter* θ, real-valued or belonging to a subset of some Euclidean space $\mathcal{R}^p$. (I will not go into nonparametric statistics in this book.) Giving θ some value defines a state of the unknown world. Look at the situation before the experiment or observational study is done, and choose some potential *observations* X_i. These observations are assumed to have a probability distribution for each given state of the world. *The specification of this class of probability distributions constitutes the statistical model.* The statistical model should focus on the relationship between the parameter that one is interested in and the observations to be done, and it should represent this relationship as well as possible. It should be simple, but not too simple.

The purpose of statistical modeling can be listed as follows:

- Give a rough description of the data generating process.
- Provide parameters that can be estimated from data.
- Allow focusing upon certain parameters.
- Give a language for asking questions about nature.
- Give means for answering such questions by estimation or by the testing of hypotheses.
- Provide confidence intervals and error estimates.
- Give a possibility to study deviations from the model and choosing new models.

Thus the model can be seen as part of a language. A model should be chosen carefully using subject matter knowledge together with a realization of what can be done statistically. If possible, the model should be scrutinized empirically. But once the model is chosen, this is an existential choice. Any conclusion is conditional, given the model.

Specifically, consider the situation with n repeated, independent observations. This is modeled by independent, identically distributed random variables $(X_1, X_2, \ldots, X_n)$ with distribution depending upon some parameter θ,

In the discrete case:

$$P_\theta(X_1 = x_1, \ldots, X_n = x_n) = \Pi_{k=1}^n p_\theta(x_k).$$

In the continuous case:

$$P_\theta(X_1 \le x_1, \ldots, X_n \le x_n) = \int_{-\infty}^{x_1} \ldots \int_{\infty}^{x_n} \Pi_{k=1}^n f_\theta(u_k) du_1 \ldots du_n.$$

Here $p_\theta(x)$ is the point probability of the individual observations and $f_\theta(x)$ is the probability density of the individual observations. For continuous models, a very common choice of the probability density $f_\theta(x)$ is the normal density (2.2). In some cases, this may be motivated by some form of the central limit theorem; in other cases it is just a matter of convenience. Here $\theta = (\xi, \sigma)$. One can distinguish between three cases: (1) σ is known and ξ is the unknown parameter. (2) ξ is known and σ is the unknown parameter. (3) Both ξ and σ are unknown. The study of statistical methods that are robust against the assumption of normality, is an active research area in statistics.

One simple discrete case is when one has n independent repeated trials, each with two possible outcomes A or A^c, often called success and failure. Assuming the same probability η of success in each trial, and letting X_k be the indicator of success in the kth trial, we have the point probabilities $p_\eta(0) = 1 - \eta$ and $p_\eta(1) = \eta$. If now Y is the number of successes in the n trials, it is a straightforward exercise to show that Y has a binomial distribution:

$$P_\eta(Y = y) = \binom{n}{y} \eta^y (1 - \eta)^{n-y}.$$

The binomial distribution gives $E(Y) = n\eta$ and $\text{Var}(Y) = n\eta(1 - \eta)$.

A multivariate generalization is the multinomial distribution: Assume n independent trials, each with k possible outcomes $A_1, \ldots, A_k$. Let $P(A_j) = \eta_j$ for each j in each trial, so that $\sum_{j=1}^k \eta_j = 1$. Let Y_j be the number of times that A_j has occurred in the n trials. Then for $\sum_{j=1}^k y_j = n$ we have

$$P(Y_1 = y_1, \ldots, Y_k = y_k) = \frac{n!}{y_1! \cdots y_k!} \eta_1^{y_1} \cdots \eta_k^{y_k}.$$

Here $\mu_j = E(Y_j) = n\eta_j$, $\text{Var}(Y_j) = n\eta_j(1 - \eta_j)$ and the covariance $\text{Cov}(Y_i, Y_j) = E(Y_i - \mu_i)(Y_j - \mu_j) = -n\eta_i\eta_j$. This can be summarized in the covariance matrix $C(Y)$ with $\text{Var}(Y_j)$ $(j = 1, \ldots, k)$ on the diagonal and $\text{Cov}(Y_i, Y_j)$ $(i, j = 1, \ldots, k)$ as off-diagonal terms.

Other standard choices of point probabilities and probability densities are listed in Chapter 1 of Lehmann (1999).

An extensive and deep article arguing for a limitation of the ordinary concept of a statistical model, is McCullagh (2002).

2.1.3 Inference for Continuous Parameters

The modern theory of statistical inference was developed by R.A. Fisher in the 1920s and the 1930s, at the same time as modern quantum theory was developed. Fisher knew about quantum theory, but did never hint at any relation to it in his own work.

From an epistemic point of view it is important in statistics to distinguish between the situation before any observations are done, and after observations are done. Before, the observations are unknown, but are modeled as stochastic variables X through the chosen statistical model. After the observations, they are known values $X = x$, and we want to use these observations to say something about the state of nature, the parameters θ. There has to be a recipe from x to the inference about θ.

The simplest concept is that of point estimation: The parameter θ is estimated by a function of the data: $\widehat{\theta}(x)$. The properties of this estimation procedure is evaluated by looking at the before-observation situation and using the statistical model: With the stochastic variable X inserted, $\widehat{\theta}(X)$ is called an *estimator*. One good property might be that the estimator is unbiased: $E(\widehat{\theta}(X)) = \theta$ or nearly so. Another good property is that it has a small variance. These two properties are sometimes combined in the requirement that the estimator should have a mean square error which is as small as possible, where

$$\text{MSE}(\widehat{\theta}(X)) = E((\widehat{\theta}(X) - \theta)^2) = \text{Var}(\widehat{\theta}(X)) + (E(\widehat{\theta}(X)) - \theta)^2.$$

A point estimator is often given together with a *standard error*: An estimate of the standard deviation of the corresponding estimator, i.e., the square root of its variance. The standard error gives an indication of uncertainty of the estimate.

In a typical before-observations situation, one has also the possibility to decide how much data one should take; this may be indexed by a number n. The simplest, but not uncommon, case is that of repeated measurements, that is, of n independent, identically distributed observations $X_n = (X_1, \ldots, X_n)$, but many more situations of this kind exist. The before-observation version of the estimation recipe, $\widehat{\theta} = \widehat{\theta}(X_n)$ is then the estimator. A weak, but desirable property of the estimator is that it should be *consistent*: $\widehat{\theta}(X_n)$ should converge in probability or almost surely to θ as n tends to infinity. A further property which is often satisfied by some central limit type theorem is that of asymptotic normality: Typically $\sqrt{n}(\widehat{\theta} - \theta)$ converges in law to $N(0, \sigma^2)$ for some variance σ^2. It is desirable that σ^2 should be as small as possible.

In fact these properties are satisfied for a large class of situations, explored by Fisher. In the case of discrete observations, the model implies joint probabilities for the data $p_\theta(x)$, in the continuous case joint probability densities $f_\theta(x)$. In both cases, when one focuses upon the θ-dependence, this function is called the likelihood $L(\theta)$. Fisher argued that one should maximize the likelihood to find good estimates of θ. The intuitive argument is that this will provide the θ-value which in a best possible way explains the obtained data. The maximum $\widehat{\theta}(x)$ is called the maximum likelihood estimate. Local extremes can be found by equating the derivative of the likelihood functions to zero. In more complicated situations one may have problems with several local extremes, but often these problems may be tackled by numerical maximization methods.

Maximum likelihood estimation is used throughout statistics in a large number of applications to a diverse set of applied sciences.

To evaluate the properties of the maximum likelihood procedure, one again turns to the pre-observation situation. Then $\widehat{\theta}(X)$ with the stochastic variable from the model inserted, is called the *maximum likelihood estimator*. For simplicity let us look at the situation with repeated independent continuous observations $X_n = (X_1, \ldots, X_n)$. Let $f_\theta(x)$ be the probability density of a single observation, and define the Fisher information by $I(\theta) = E((\frac{\partial}{\partial \theta} \ln f_\theta(x))^2)$ assuming that this exists. Then under regularity conditions (see for instance Lehmann 1999, where uniqueness of the local extrema is assumed for this), one can prove the following: (1) $\widehat{\theta} = \widehat{\theta}(X_n)$ is consistent; (2) $\sqrt{n}(\widehat{\theta} - \theta_0)$ converges in law to $N(0, 1/I(\theta_0))$ under θ_0, the true value, as $n \to \infty$. Thus the maximum likelihood estimator has some good asymptotic properties, and these results may be generalized to other cases with a large amount of data. However, there exist many examples of cases where the maximum likelihood estimator does *not* behave in an optimal way; see Le Cam (1990).

A good estimator $\delta(X)$ can also be found using a loss function $L(\delta(X), \theta)$, for instance quadratic loss $L = (\delta(X) - \theta)^2$. One objective might be to minimize the risk, or expected loss, $R(\delta, \theta) = E_\theta(L(\delta(X), \theta))$, but a uniform minimization here is not feasible: Taking $\delta(X) \equiv \theta_0$ gives $R(\delta, \theta_0) = 0$ for all reasonable loss functions. One way around this, is to limit oneself to unbiased estimators: $E_\theta(\delta(X)) = \theta$ for all θ. The theory on this can be found in Lehmann and Casella (1998).

In addition to point estimation, statistical inference theory discusses hypothesis testing and confidence intervals. Hypothesis testing is closely related to confidence intervals. I will consider here one-sided confidence intervals $(-\infty, \bar{\theta}]$ and two-sided confidence intervals $[\underline{\theta}, \bar{\theta}]$. The lower and upper limits of these intervals are functions of the data. When considered again in the pre-observational situation, they should have the properties

$$P_\theta(\theta \in (-\infty, \bar{\theta}(X)]) = P_\theta(\bar{\theta}(X) \geq \theta) = \gamma, \qquad (2.3)$$

$$P_\theta(\theta \in [\underline{\theta}(X), \bar{\theta}(X)]) = P_\theta(\underline{\theta}(X) \leq \theta \leq \bar{\theta}(X)) = \gamma, \qquad (2.4)$$

where γ is a pre-assigned confidence coefficient, say 0.95 or 0.99.

The statistical methods discussed so far are called frequentist methods: They are coupled to a pre-observational distribution using the statistical model. The probabilities and expectations involved in this can be interpreted by a thought construction: Imagine that the whole experiment is repeated a large number of times. Then the imagined relative frequency of an event A for these repetitions is approximately equal to the hypothetical probability $P(A)$. The probabilities and expectations are therefore connected to the methods used and to the statistical model used.

There is another approach to statistical inference which has a long history, but has been particularly popular in the last few years: The Bayesian approach. Here the probabilities are imagined to be connected to the parameters themselves. The important assumption is that one first in some way has obtained a prior distribution on the parameter, say with a probability density $\pi(\theta)$. From this prior, one finds a posterior distribution, given the data, by using a variant of Bayes' formula

$$P(T|D) = \frac{P(T \cap D)}{P(D)} = \frac{P(T)P(D|T)}{P(D)}.$$

The first part of this formula is the definition of the conditional probability of T, given D. This definition is consistent with the Radon-Nikodym approach, and also consistent with what one calls conditional probability in simple examples. The second part of the formula is a consequence. Applied to a situation with a continuous parameter θ and a continuous data model with density $f_\theta(x)$, a formula for the posterior density of θ given the data is obtained:

$$\pi(\theta|x) = \frac{\pi(\theta)f_\theta(x)}{f(x)} = \frac{\pi(\theta)f_\theta(x)}{\int \pi(\phi)f_\phi(x)d\phi}.$$

Consider first Bayesian point estimation. Again defining a loss function $L(\delta(x), \theta)$, we can now introduce the Bayesian risk as

$$BR(\delta(x)) = \int L(\delta(x), \theta)\pi(\theta|x)d\theta,$$

and find the estimate $\delta(x)$ which minimized BR. With quadratic loss this leads to the posterior mean $\int \theta\pi(\theta|x)d\theta$ as an estimate. Other possible estimates include the mode and the median of the posterior distribution, the mode being the maximum of the density and the median is the value such that the probability that the parameter is below this value, equals 1/2. In these estimates one can insert the pre-observational stochastic variable X, compare them with estimators obtained by frequentists methods, evaluating estimators using a frequentist or Bayesian approach.

The Bayesian concept which replaces the confidence intervals is that of credibility intervals. Again consider the one-sided case $(-\infty, \theta^*(x)]$ and the two-sided case $[\theta_*(x), \theta^*(x)]$. These intervals have direct interpretations in terms of a probability distribution over the parameter, the posterior distribution:

$$P(\theta \in (-\infty, \theta^*(x)]) = \int_{-\infty}^{\theta^*(x)} \pi(\theta|x)d\theta.$$

$$P(\theta \in [\theta_*(x), \theta^*(x)]) = \int_{\theta_*(x)}^{\theta^*(x)} \pi(\theta|x)d\theta.$$

To choose $\theta_*(x)$ and $\theta^*(x)$, this can again be given a preassigned value γ, say 0.95 or 0.99. In a specific sense, the interpretation of the credibility interval is simpler and more direct than the interpretation of the confidence interval.

There is much more to say about Bayesian theory and Bayesian methods; see Bernardo and Smith (1994), Box and Tiao (1973) and Congdon (2006).

The great weakness with the Bayesian approach is that the scientist should be able to specify a prior distribution of the unknown parameter. In a way he should be willing to and able to enter a let on the values of this parameter. It is often claimed that if the scientist is not willing to do this, he should use an objective prior; for different formal ways to specify this concept, see Kass and Wasserman (1996). I have recently used such a prior myself (see Helland et al. 2012), but even so I would claim: There are many cases where the scientist could not or should not have any fully specified prior opinion about the parameter, even not one based upon symmetry or other 'objective' criteria. In such cases he should resort to frequentist methods. In statistical inference one should be flexible, not staying with one approach which should be imagined to cover all cases. Many modern statisticians argue for such a flexible attitude. An example is Efron (2015), who proposes a way to measure the frequentist accuracy of Bayesian estimates.

Quite recently there has been proposed a frequentist alternative to a distribution connected to a parameter: The confidence distribution; see Schweder and Hjort (2002, 2016) and Xie and Singh (2013). The main idea is that one looks upon the confidence interval for any value of the confidence coefficient γ. Let $(-\infty, \tau(\gamma, x)]$ be a one-sided confidence interval with coefficient γ, where $\tau(\gamma) = \tau(\gamma, x)$ is an increasing function. Then $H(\cdot) = \tau^{-1}(\cdot)$ is the confidence distribution for θ. This H is a distribution function and has the property that $H(\theta, X)$ has a uniform distribution over the interval $[0, 1]$ under the model.

In general, look at any one-dimensional parameter θ, a function of the total parameter ϕ. Then $H(\theta, X)$ is the cumulative distribution for a confidence distribution for θ provided that it has a uniform distribution under the distribution of X, whatever the true value of ϕ.

Confidence distributions are often found by means of a *pivot*. A function $piv(\theta, X)$ is a pivot it its distribution function $G(u) = P(piv(\theta, X) \leq u)$ is independent of the underlying parameter ϕ. If the pivot is increasing in θ, we can take $H(\theta, X) = G(piv(\theta, X))$.

According to Xie and Singh (2013), and also Schweder and Hjort (2016), which is a very good monograph over confidence distributions, the distribution function H is to be looked upon as a distribution *for* the parameter, to be used in the epistemic process, not a distribution *of* the parameter, as we have in the Bayesian approach. Xie and Singh (2013) use the confidence distribution as a general concept in order to build a bridge between the Bayesian approach and various interpretations of the frequentist approach. This may be seen to be in the spirit of the present book.

A confidence probability for a parameter is in Schweder and Hjort (2016) called an epistemic probability, and is of a different nature than ordinary probabilities, called *aleatoric* probabilities. The latter are coupled to real randomness in the world, and not only to our lack of knowledge attached to phenomena in the world. The confidence distributions are coupled to epistemic probabilities, or as Schweder and Hjort write, they express what a rational observer thinks about the world in the light of the model and the data. Schweder and Hjort claim that the confidence distributions are objective, in contrast to most Bayesian aposteriori distributions.

The following is a straightforward translation from a book report by Hermansen et al. (2017):

'For many Bayesians this distinction between two different interpretations of probability is well known. For example Gelman and Robert (2013) write that "priors are not reflections of a hidden "truth" but rather evaluation of the modeler's uncertainty about the parameter." For the Bayesians of this school the parameter θ is a fixed, but unknown quantity, the prior distribution $\pi(\theta)$ is epistemic, while $f(y|\theta)$ is aleatoric... One might say that one as Bayesian not is forced to regard θ as a random variable from an *ontological* point of view, only mathematical. Schweder and Hjort's claim that the Bayesian "has only one form of probability, and has no choice but to regard parameters as stochastic variables", is therefore a simplification. It is true that the Bayesian only has one form for probability theory, the one built upon Kolmogorov's axioms, but that does not mean that the Bayesian is forced to have one interpretation of what probability *is*.

While Bayesians like Gelman and Robert have one probability theory and two interpretations of probability, the confidencialist is forced to have two probability theories, the one we know for the aleatoric and something else for the epistemic. This is because the confidence probabilities violate Kolmogorov's axioms.'

In the general language of the present book one might say that statisticians are divided into three complementary schools, Bayesians, Frequentists and Fiducials, where the last school is inspired by what some call Fisher's largest blunders, but which contains techniques that has gained new popularity in recent years. It is not quite true that these schools are completely disjoint; good statisticians may use techniques from two or three schools. One can also find connections between the various approaches. There are now being arranged yearly international joint Bayesian, Frequentist and Fiducial (BFF) conferences (informally: Best Friends Forever), where both practical, technical and philosophical questions are discussed.

Three general book on statistical inference are Casella and Berger (1990), Bickel and Doksum (2001) and Cox (2006). For a discussion of Fisher's contributions with a view towards the future, see Efron (1998). The recent book by Cox and Donnelly (2011) discusses many aspects of applied statistics and also provides some links to theoretical statistics.

2.1.4 Inference for Discrete e-Variables

Opinion polls, or sample surveys, while very much used in practice, are not much discussed in the standard mathematical statistical literature. But specialized books like Cochran (1977) exist. The framework is that one has a finite population consisting of N units, say N human beings, and one seeks some information about this population from a smaller sample of n units. The simplest case is when the sample is taken randomly from a register of the whole population, but other sampling plans exist. To increase efficiency, one often uses stratified sampling: The population is divided into k strata using some relevant criterion, such that the numbers of unit in stratum i is N_i, and one samples randomly n_i units from this stratum. Of course $\sum_{i=1}^{k} N_i = N$ and $\sum_{i=1}^{k} n_i = n$.

As a simple epistemic problem, assume that an unknown number M in the population possesses some specific property A, and one wants to use the sample to estimate $\theta = M/N$. This θ takes a discrete set of values $0, 1/N, \ldots, N/N$, and is not always called a parameter. In this book we will use the more general concept of an e-variable, a conceptual variable which is unknown before the epistemic process begins. In general it is implicit in the concept of an e-variable that this is a quantity that we want to gain knowledge about.

A more general problem is that each unit j in the population has some value y_j attached to it, and one wants to estimate $\theta = \bar{y}_{population} = \sum_{j=1}^{N} y_j/N$. The simple problem above is then obtained by specializing y_j to be an indicator function. A common estimate of θ is

$$\hat{\theta} = \frac{\sum_{sample} y_j/\pi_j}{\sum_{sample} 1/\pi_j},$$

where π_j is the probability that unit j should be included in the sample. This can be used for many sampling plans. For stratified sampling we get $\hat{\theta} = \sum_{i=1}^{k} N_i \bar{y}_i/N$, where the mean $\bar{y}_i$ is over the sample in stratum i.

Opinion polls are based on the assumption that each person 'has' an opinion on the issue that is focused upon. The fact that opinions may vary with time, and that they may depend on the contexts, is perhaps realized, but it is not always discussed in this connection. To see this first from the point of view of the person being interviewed, imagine for instance that a woman A has spent some time on an hotel, and then after a few days receives a questionnaire by e-mail, one of the questions being: 'On a scale from 1 to 10, how do you evaluate the service at this hotel?' This causes her to enter an epistemic process, mostly related to introspection. To begin with, the score is some unknown number θ, but after a while she decides on a value $\theta = u_k$, one of the values from 1 to 10.

This decision process may be evaluated subjectively by the woman A herself. We may also consider the whole situation as looked upon by some person B knowing her background. In this latter case prediction is relevant. The person B may have access to some kind of data from a sample of size n_i from a stratum consisting of people with the same background as the woman, and to detailed information about the hotel. On this basis he may want to predict θ. Again this is an epistemic process with a discrete e-variable; the target for the prediction of this e-variable is not a population, but a single unit, the woman A in this case. B may wish a large n_i to have accurate data, but at the same time resources may be limited: He may be forced to have a small n_i in order to be able to predict from a fairly homogeneous subpopulation.

Both in the introspection case and in the external observer case, one can consider the following: Assume that the woman A had an unfortunate episode with the receptionist of the hotel just before she left, and that the observer B does not know about this episode. Let θ' be the hypothetical score that A would have had if the episode had not taken place. Depending on the further circumstances, θ' may not be accessible to the woman A herself, and θ may not be accessible to the observer B.

Considerations of this kind are here very vague, but they should give some feeling of what I mean by an epistemic process when a single unit is involved. The situation here is on the borderline or outside what one meets in ordinary statistics, but the point is that it describes epistemic processes, and that one of these processes (the prediction part) in principle can be made precise in statistical terms.

Such considerations will be important when I later will try to approach the foundation of quantum mechanics from an intuitive point of view. The variables of interest in quantum mechanics are discrete or continuous e-variables. They have some properties in common with the parameters of statistics, and some properties in common with the example just sketched. They are connected to single units (particles) or to a few units. It is important to realize, however, that the e-variables of quantum mechanics are connected to the microworld and to the relevant observers. They have properties analogous to the e-variables of the macro-world, but they must be considered as unique concepts.

The Bayesian concepts of prior and posterior distribution are straightforward to formulate in the case of a discrete e-variable, and the concept of confidence distribution also carries over: If θ takes the values $u_1, \ldots, u_r$, then the confidence coefficient γ can take only r values, and the confidence distribution H is determined as follows: Let again $(-\infty, \tau(\gamma, x)]$ be a one-sided confidence interval with coefficient γ, where $\tau(\gamma) = \tau(\gamma, x)$ on the r values. Then $H(\cdot) = \tau^{-1}(\cdot)$ can be extended to a discrete distribution function for θ, which has the property that $H(\theta, X)$ for data X has a uniform distribution on the r values $H(u_1), H(u_2), \ldots, H(u_r) = 1$. Strictly speaking, this does not fit into the general definition of Schweder and Hjort (2016) of a confidence distribution, which requires a uniform distribution on the unit interval, but it satisfies their definition of an asymptotic confidence distribution. Since discrete e-variables will play an important role in the last part of the book, I will take the liberty to use the word confidence distribution here also.

2.2 Group Actions and Model Reduction

In simple random sampling, a natural objective prior for the e-variable θ is found by giving the same probability $1/N$ to each unit in the population. This is the invariant measure (see Appendix B) for the permutation group. In general an epistemic problem related to a θ may often have some symmetry property associated with it, and this is formalized by introducing a group of transformations acting upon the space Θ of the e-variable. When θ is transformed by the group and the observations are transformed accordingly (see Helland 2004), one should get equivalent results from the statistical analysis. As a trivial example: One should get equivalent results from a statistical analysis whether the parameters and the observations are measured in meters or in centimeters. A summary of basic group theory is given in Appendix B. Examples of groups acting upon a parameter space, are location: $\xi \to \xi + a$ for a real; scale group: $\sigma \to b\sigma$ for $b > 0$; location and scale: $(\xi, \sigma) \to (a + b\xi, b\sigma)$, where ξ is an expectation and σ is a standard deviation; rotation in a multidimensional parameter space; a general linear group acting upon a multidimensional parameter space etc.. Invariance under a group may help improving the estimation or the inference in general.

I will not be very precise on the choice of G, but just say vaguely that we choose G, if possible, in agreement with some symmetry aspect of the whole situation.

The model reduction that I will introduce now, will later be relevant also in a quantum theory setting. In order to be general, I will now for a moment use the term 'e-variable' instead of 'parameter'. For the present discussion one can just think of this as a parameter, however.

Now fix a point θ_0 in the e-variable space Θ. An *orbit* in this space under G is the set of points of the form $g\theta_0$ as g varies over the group G. The different orbits are disjoint, and θ_0 can be replaced by any e-variable on the orbit. Any set in Θ which is an orbit of G or can be written as a union of orbits, is an invariant set under G in Θ, and conversely, all invariant sets can be written in this way. If there is only one orbit in Θ, the group is said to be acting transitively upon Θ.

A statistical model should be as simple as possible, but not simpler. In some cases we may want to do a simplification, a model reduction. This may take the form of a reduction of the e-variable space Θ. Parts of this space which are essential for the epistemic process, must always be retained, but irrelevant dimensions should be left out. I will now formulate a general criterion which will be used throughout this book:

If there is a group G acting upon the e-variable space Θ, any model reduction should be to an orbit or to a set of orbits of G.

This will ensure that G also can be seen as a group acting upon the new e-variable space. In particular, if the group actions form a transitive group G, no model reduction is possible.

Example 2.1 Assume that a single set of observations is modeled by some large parametric model, only assuming that parametric class contains the normal model. Let the location and scale group be acting upon the parameter space Θ. Then one orbit is given by the $N(\xi, \sigma^2)$ distribution. This is not an uncommon model reduction.

Example 2.2 Look at two independent sets of observations: $(X_1, \ldots, X_m)$ independent and identically $N(\xi_1, \sigma_1^2)$ and $(Y_1, \ldots, Y_n)$ independent and identically $N(\xi_2, \sigma_2^2)$. Let G be the translation and scale group given by $\xi_1 \to a_1 + b\xi_1$, $\sigma_1 \to b\sigma_1$, $\xi_2 \to a_2 + b\xi_2$, $\sigma_2 \to b\sigma_2$. Note that a common scale transformation by b is assumed. Then the orbits of the group in the parameter space are given by $\sigma_1/\sigma_2 = $ constant. A common model reduction is given by $\sigma_1 = \sigma_2$. This simplifies the comparison of ξ_1 and ξ_2, which is often the goal of the investigation.

Example 2.3 Linear statistical models have a large range of applications. In general these models have the form where the observations Y_l are independent $N(\xi_l, \sigma^2)$, where the expectations ξ_l are linear combination of a set of parameters. In one particular such model (the two-way analysis of variance model) the observations Y_{ijh} have expectations $\mu + \alpha_i + \beta_j + \gamma_{ij}$. To get a unique representation of this kind, one often imposes the restrictions $\sum_i \alpha_i = 0$, $\sum_j \beta_j = 0$, $\sum_i \gamma_{ij} = 0$ for each j and $\sum_j \gamma_{ij} = 0$ for each i. Let the group G be given by all permutations of the index i and all permutations of the index j. Then an obvious model reduction is given by the invariant set where the expectation is $\mu + \alpha_i + \beta_j$. This is called the model without interaction, and is a valid simplification in some cases.

Example 2.4 Another example of a linear model is the polynomial regression model $Y_i = \beta_0 + \beta_1 x_i + \ldots + \beta_p x_i^p + E_i$, where the E_i's are independent $N(0, \sigma^2)$ for $i = 1, \ldots, n$. Let G be the group defined by translations in the x-space: $x \to x + a$, which generates a transformation group on the parameters $(\beta_0, \ldots, \beta_p)$. Then the submodels $Y_i = \beta_0 + \beta_1 x_i + \ldots + \beta_q x_i^q + E_i$ $q < p$ correspond to invariant sets in the parameter space.

Example 2.5 A further example of a linear model is the multiple regression model $Y_i = \beta_0 + \beta_1 x_{i1} + \ldots + \beta_p x_{ip} + E_i$ for $i = 1, .., n$ with fixed x_{ij}, which again has many different applications. Consider first the case where the x_{ij} are measured in different units for different j. Then there is a natural transformation group given by separate scale changes $x_{ij} \to k_j x_{ij}$ $(j = 1, \ldots, p)$. This induces a group on the regression parameters by $\beta_j \to \beta_j / k_j$ $(j = 1, \ldots, p)$. The invariant sets in the parameter space are found by putting some of the β_j's equal to 0. These reduced models are well-known from many applications of regression analysis.

Example 2.6 Consider the same multiple regression model as in Example 2.5, but assume now that the explanatory variables x_{ij} all are measured in the same units. A large class of transformations $x_{i.} \to Q x_{i.}$ may then be of interest. In particular, an interesting case is when Q varies over the orthogonal matrices.

As here, and as in any linear model, estimates of the regression parameters can in principle be found by the method of least squares, which is equivalent to the maximum likelihood method. However, this method breaks down when $p > n$, or more generally when one has collinearity problems such that the matrix which we need to invert in order to implement the least squares solution, is singular. A large number of alternative estimation methods are proposed in the statistical literature to tackle this problem, but it seems very difficult to decide which of these methods one should use in practice.

For this problem, one place where one may start the investigation, is that many of the methods are *equivariant* under the transformation induced by rotation in the x-space: A transformation on $\widehat{\theta}$ found from transformations of the data is the same the corresponding transformation on the parameter θ.

Before I return to this problem, I will summarize a little more theory. In Appendix B the concept of a right invariant measure for the group is defined, and it is recommended that such a prior is used as an objective prior. Among other things it is proved in Helland (2004, 2010) that there for a transitive group is a very close connection between confidence intervals and Bayesian credibility intervals in this case. It follows from this that there is a close connection between confidence distributions and posterior distributions with this prior.

Concerning equivariant estimators, there is a generalization of an old theorem by Pitman, which is proved in Helland (2010), showing that if the loss function is invariant and proportional to the quadratic loss, if the group is transitive and a right invariant prior is used, then the posterior mean, if finite, is the best equivariant estimator.

Example 2.6, continued   Look at a modification of the model in Example 2.6 where the explanatory variables are random variables X_{ij}. This is natural in many observational studies. For simplicity, assume that all variables are centered: $E(Y_i) = 0$ and $E(X_{ij}) = 0$. Then the model is $Y_i = \beta_1 X_{i1} + \ldots + \beta_p X_{ip} + E_i$ for $i = 1, .., n$. Let Σ_x be the covariance matrix of the x-variables, which can be defined by the property that $\text{Var}(\sum_j a_j X_{ij}) = a^T \Sigma_x a$ for all vectors $a = (a_1, \ldots, a_p)^T$. Then $\beta = (\beta_1, \ldots, \beta_p)^T$ can always be expanded in terms of an orthogonal set of eigenvectors d_i of Σ_x:

$$\beta = \sum_{i=1}^{p} \gamma_i d_i. \tag{2.5}$$

In this expansion the number of terms can be reduced in two ways: (1) Some of the eigenvalues may be coinciding. Then the eigenvectors in this eigenspace can be rotated in such a way that there is just one eigenvector in this space which has a nonzero component along β. (2) The vector β has no component in this eigenspace. So an interesting reduced model is the one with m non-zero terms in (2.5). The ordering of the terms in (2.5) is arbitrary, so the reduced models only specify the number m of non-zero terms, not which terms that are non-zero. It is not difficult to show that these models for different m are exactly the orbits of the following group G: Rotations in the x-space and hence of the eigenvectors d_i augmented by independent scale transformations $\gamma_i \rightarrow a_i \gamma_i$ where $a_i > 0$.

It is shown in Helland et al. (2012) and Cook et al. (2013) that these reduced models coincide with reduced models introduced by researchers from two different traditions: The envelope model of Cook et al. (2010) and a natural population model arising from the partial least squares algorithmic 'soft' models from chemometrics. Maximum likelihood estimation and other estimators under the reduced model are discussed in Cook et al. (2013) and Bayes estimation in Helland et al. (2012). The invariant prior induced by the group leads to an undefined posterior expectation, so a best equivariant estimator can not be found from this. However approximating the scale prior with a proper prior leads to β-estimates, hence predictions, which seem to have good properties.

These two estimators assume a multinormal distribution of $\{X_{ij}, Y_i\}$. However, both the chemometric tradition and the envelope model of Cook et al. (2010, 2013) demand no detailed distributional assumptions.

A systematic comparison of estimators under the multinormal envelope/PLS model is now carried out by Helland et al. (2018). This is based upon a computer package which is being developed by Sæbø et al. (2015).

Finally I give for completeness a simple example of model reduction for the case where the e-variable in question is not a continuous parameter.

Example 2.7 In stratified random sampling, the natural group G is the group of independent permutations within each stratum. The orbits of G are then given by the single strata, and model reductions to invariant sets are given by reduction to any set of strata. Such a reduction is of course natural in cases where one want to limit the investigation to a particular set of strata.

2.3 Interlude

Before the epistemic process, one has an unknown e-variable θ. What is the situation after one has gone through an epistemic process? In the case where θ is the parameter of a statistical model, the situation can be summarized as follows: Depending upon the statistical philosophy used, one has either a confidence interval $[\underline{\theta}(x), \bar{\theta}(x)]$ or a credibility interval $[\theta_*(x), \theta^*(x)]$. In both cases, assume that the coefficient γ is very high, say 0.999. Then the practical conclusion from the epistemic process is that the new state is specified by saying that θ belongs to this interval.

Confidence intervals or credibility intervals for discrete e-variables are not much discussed in the statistical literature, but the concepts carry over. One difference is that when one has very much data or the experimental uncertainty is very low, the intervals can degenerate into a single point. In the following, it will make the discussion much simpler to consider such a case. Assume that one has this situation, and let again the coefficient γ is very high, say larger than 0.999. Then in the frequentist case, one has a conclusion of the type $P_\theta[\widehat{\theta}(X) = u_k] = \gamma$ with realized data $X = x$, and in the Bayesian case one has a posterior probability $P[\theta = u_k | x] = \gamma$. In both cases we conclude for practical purposes that the new state is given by $\theta = u_k$, and that this value can be used in further investigations.

Any epistemic process starts with an unknown e-variable θ, and when the process ends, one has some knowledge about θ. A state is obtained when this knowledge is almost certain. In the simplest case the knowledge can be expressed by a certain fixed value $\theta = u_k$. This situation can be realized by a statistical investigation with a discrete parameter/e-variable, but it can also be realized in other epistemic situations. One example is when a person through introspection makes up his or her mind on a particular issue, as illustrated with the woman answering an opinion poll in Sect. 2.1.4. Again we can talk about a state when the person's knowledge about his/her opinion is almost certain. Looking upon the process of achieving an opinion on an issue as an epistemic process is an example of a *decision*. From a more qualitative point of view decisions are discussed in Sects. 3.3, 6.1 and 6.2 below. Many examples of everyday decisions can be given, some realized through the communication with other people. Some of these processes start with an unknown θ and end up with an (almost) sure state $\theta = u_k$. Other examples of this are connected to prediction of some variable. In these last examples, the e-variable is typically attached to a single unit, not to a population of units, which is the most common situation in statistical investigations.

The example in Sect. 2.1.4 illustrates another issue: Here one e-variable θ is accessible to the woman A herself, while another e-variable θ' (the hypothetical score if a certain episode had not taken place) is accessible to the person B knowing her background and having information about the hotel. The reason for this difference is that the two persons have different background knowledge. We will come back to similar situations later when discussing quantum mechanics.

Returning now to statistics, nearly all papers on statistical inference have data models, that is, either parametric or nonparametric models of the observed data, as their point of departure. Also, statistical practice is deeply founded upon this tradition. Even though a different culture was promoted and discussed by Breiman (2001), the data modeling culture is now more dominant than ever.

Nor will this book depart radically from this culture, but we will add an element to it: Every decision in any experimental or observational setting is made in a context. This context may not be trivial, and may have decisive influence on how the inference should be made. The context may be in parts be formed by the historical and cultural background for the study, and it may depend upon earlier decisions. But it can also in addition be conceptual, including the formulated goal of the investigation, the model, a loss function and/or a Bayesian prior. Also the framework for the study must be considered as a part of the context: The experimental units available, what can be measured on these units; limitations in terms of money, time and human resources.

In order to be able to discuss contexts in general, it turns out to be useful first to give a precise definition of what we mean by conceptual variables, which includes observations, parameters, latent variables and more. Then we define e-variables, which up to now has been a loosely defined concept.

In statistical theory, a parameter is often defined as an index of a class of distributions, but in statistical practice, a parameter is often a quantity of interest in itself, introduced as an expectation, a variance, a covariance, a correlation, a regression coefficient or a probability. These two facets of the parameter concepts may to some extent be regarded as complementary, even though this introduces no logical difficulty.

In the statistical tradition, a parameter is usually connected to a hypothetical infinite population, but in fact parts of the statistical theory - in reality nearly everything that is not related to asymptotical considerations - can be generalized to the case where the parameter is replaced by a conceptual variable connected to a single unit or to a few units.

In addition to the unknown conceptual variable there are data z. The purpose of an experiment is then to use these data to answer questions formulated in terms of the conceptual variable. This will be the background for our approach to essential parts of quantum theory later in this book. Quantum theory is not statistics, but both give examples of epistemic processes involving observations.

The present book concentrates upon estimation and prediction, but the conceptual framework discussed here is also valid for other types of statistical inference. Confidence intervals may be considered if the context contains a set of hypothetical situations where a particular estimation procedure is used, and Bayesian analysis is relevant if prior distributions of parameters are part of the context. Finally, in the case of a hypothesis testing setting, which is not much discussed in this book, the context may contain a specification of a null hypothesis and an alternative hypothesis. Or if we are interested in Fisherian p-value testing, a null hypothesis and a direction for the alternative should be specified in the context.

References

Bernardo, J. M., & Smith, A. F. M. (1994). *Bayesian theory.* Chichester: Wiley.

Bickel, P. J., & Doksum, K. A. (2001). *Mathematical statistics. Basic ideas and selected topics* (2nd ed.). New Jersey: Prentice Hall.

Box, G. E. P., & Tiao, G. C. (1973). *Bayesian inference in statistical analysis.* New York: Wiley.

Breiman, L. (2001). Statistical modeling: The two cultures. *Statistical Science, 16*, 199–231.

Casella, G., & Berger, R. L. (1990). *Statistical inference.* Pacific Grove, CA: Wadsworth and Brooks.

Cochran, W. G. (1977). *Sampling techniques* (3rd ed.). New York: Wiley.

Congdon, P. (2006). *Bayesian statistical modelling* (2nd ed.). Chichester: Wiley.

Cook, R. D., Li, B., & Chiaromonte, F. (2010). Envelope models for parsimonious and efficient multivariate linear regression. *Statistica Sinica, 20*, 927–1010.

Cook, R. D., Helland, I. S., & Su, Z. (2013). Envelopes and partial least squares regression. *Journal of the Royal Statistical Society, Series B, 75*, 851–877.

Cox, D. R. (2006). *Principles of statistical inference.* Cambridge: Cambridge University Press.

Cox, D. R., & Donnelly, C. A. (2011). *Principles of applied statistics.* Cambridge: Cambridge University Press.

Efron, B. (1998). R.A. Fisher in the 21st century. *Statistical Science, 13*, 95–122.

Efron, B. (2015). Frequency accuracy of Bayesian estimates. *Journal of the Royal Statistical Society, B, 77*, 617–646.

Gelman, A., & Robert, C. P. (2013). "Not only defended but also applied": The perceived absurdity of Bayesian inference. *The American Statistician, 67*, 1–5.

Helland, I. S. (2004). Statistical inference under symmetry. *International Statistical Review, 72*, 409–422.

Helland, I. S. (2010). *Steps towards a unified basis for scientific models and methods.* Singapore: World Scientific.

Helland, I. S., Sæbø, S., & Tjelmeland, H. (2012). Near optimal prediction from relevant components. *Scandinavian Journal of Statistics, 39*, 695–713.

Helland, I. S., Sæbø, S., Almøy, T., & Rimal, R. (2018). Model and estimators for partial least squares regression. *Journal of Chemometrics.* https://doi.org/10.1002/cem.3044.

Hermansen, G., Cunen, C., & Stoltenberg, E. A. (2017). *Ny bok: Confidence, likelihood, probability. statistical inference with confidence distributions.* (In Norwegian). *Tilfeldig Gang, 34* no. 1.

Kass, R. E., & Wasserman, L. (1996). The selection of prior distributions by formal rules. *Journal of the American Statistical Association, 91,* 1343–1370.

LeCam, L. (1990). Maximum likelihood: An introduction. *International Statistical Review, 58,* 153–171.

Lehmann, E. L. (1999). *Elements of large-sample theory.* New York: Springer.

Lehmann, E. L., & Casella, G. (1998). *Theory of point estimation.* New York: Springer.

McCullagh, P. (2002). What is a statistical model? *Annals of Statistics, 30,* 1225–1310.

Schweder, T., & Hjort, N. L. (2002). Confidence and likelihood. *Scandinavian Journal of Statistics, 29,* 309–332.

Schweder, T., & Hjort, N. L. (2016). *Confidence, likelihood, probability. Statistical inference with confidence distributions.* Cambridge: Cambridge University Press.

Sen, P. K., & Singer, J. M. (1993). *Large sample methods in statistics.* London: Chapman and Hall, Inc.

Sæbø, S., Almøy, T., & Helland, I. S. (2015). Simrel - a versatile tool for linear model data simulation based on the concept of a relevant subspace and relevant predictors. *Chemometrics and Intelligent Laboratory Systems, 146,* 128–135.

Xie, M., & Singh, K. (2013). Confidence distributions, the frequentist distribution estimator of a parameter - a review. Including discussion. *International Statistical Review, 81,* 1–77.

Chapter 3
Inference in an Epistemic Process

Abstract A conceptual variable is defined, and on this background the e-variable is given a precise definition. Also, the context of an epistemic process is defined. One can formulate a precise setting called a generalized experiment, and in this setting sufficiency and conditioning are discussed. The conditionality principle and the sufficiency principle are seen as intuitively obvious. Birnbaum's theorem states that the likelihood principle follows from these two principles. Everything is seen in the context of an epistemic process.

3.1 Conceptual Variables and Contexts

Fisher (1922) introduced the concept of parametric models in the way it is used throughout statistics today. According to Stigler (1976) and Cook (2007), the word 'parameter' is mentioned 57 times in that groundbreaking paper. Recently, Taraldsen and Lindqvist (2010) argued that in Bayesian inference the parameters and the potential observations should be defined on the same underlying measure space. This is one point of departure of the present book. Another point of departure is that in any situation where inference is supposed to be done, several other types of unknown variables than parameters are of relevance (one simple example is in prediction), and that additional types of variables are needed to describe the context of the experiment or observational study.

Definition 3.1 Consider any experimental or observational situation at a given time or over some time span, more generally any epistemic process. Any variable which can be defined in words by a person or by a group of persons in that situation is called a conceptual variable.

This term was indicated in Helland (2010), where it also was argued that some unknown conceptual variables could be inaccessible, that is, they could not be given a value with arbitrary accuracy through estimation or prediction in any way in the given situation. This was taken as the first steps in a line of reasoning indicating a connection between theoretical statistics and quantum theory, a line of reasoning that I will continue below.

© Springer-Verlag GmbH Germany, part of Springer Nature 2018
I. S. Helland, *Epistemic Processes*, https://doi.org/10.1007/978-3-319-95068-6_3

In the following, I may alternately speak about one conceptual variable and several related conceptual variables in the same way as we may talk about a multivariate parameter or several one-dimensional parameters.

Several classes of conceptual variables are of interest in a statistical investigation, depending upon the situation:

- Context variables: The background variables for an experiment or observational study.
- OCV's (observed conceptual variables): Data or preset values.
- Statistics: Known functions of the data.
- Quasi-statistics: Known or unknown functions of the data.
- Input variables and responses/output variables. As used in prediction and regression, cp. Hastie et al. 2009.
- UCV's (unknown conceptual variables): For instance parameters, latent variables or a response for a new set of input variables.
- Hypothesis variables: Concepts from which one may formulate assertions about the value of a parameter.
- Conclusion variables: Conceptual variables by which one may formulate the conclusions from an experiment or observational study.

  Most variables in this book are either real variables, vector variables or variables belonging to some function space. But in principle a variable can be anything. The conclusion variable may be a journal article, its summary or part of the article.

In this book I will consider any epistemic process.

Definition 3.2 A conceptual variable which is used in an essential way in an epistemic process is called an *e-variable* θ. Before the epistemic process is started, the e-variable is unknown. After the process, one is able to achieve some conclusion about the e-variable, the simplest case being that we know its value: $\theta = u_k$, a type of conclusion which is only possible for discrete e-variables.

In this book the concept of e-variable will be very important. In the same way as a parameter in statistics can be any property for a population, an e-variable can represent any property for a population, a single unit or a group of units. The sole requirement is that is shall be used in an epistemic process. Like a parameter, it is a theoretical variable before the process, but one will have concrete information about the value of the e-variable after the epistemic process.

As indicated in Chap. 1, the concept of e-variables is also important in quantum mechanics. In this and the following chapters I will focus much on discrete simple e-variables, which are essentially what is called observables in quantum mechanics. The fact that these observables may be associated with operators of a Hilbert space, will be the theme of the next chapters. The discussion of the present chapter will focus upon generalized experiments and upon properties that are shared between the parameters of experiments in statistics and the e-variables of experiments in quantum mechanics. We have similar kinds of epistemic processes in mind for the two cases.

We also have the important concept of a *context* of an epistemic process.

Definition 3.3 In the case of an experiment the context includes the setting of the experiment, similarly for an observational study. But in general for any epistemic process it also includes the background for the process, historical, specific and conceptual. The conceptual background for any study should always include a formulated goal of the study.

In a series of experiments or in a meta-analysis, the conclusions from one situation may be used as a part of the context of the next situation. In a quantum mechanical situation, it is important to distinguish between the preparation of the physical system and the measurement on the system. The preparation then forms the context for the measurement.

Several operations may be done on any assertion containing conceptual variables, including $\neg$ (negation), $\wedge$ (and) and $\vee$ (or). Formulating statements connected to a concrete experimental or observational situation may then be done using propositional logic, a subject which has a large abstract literature. As formulated in Appendix E, I want to be more concrete and regard sentences formulated in ordinary, everyday language as primitive entities.

There is a close connection between propositional logic and set theory, where we identify $\neg$ with complement, $\wedge$ with intersection and $\vee$ with union. Such identifications are often done implicitly in elementary textbooks in probability. Let $(\Omega', \mathcal{F}')$ be the measurable space thus obtained, where $\mathcal{F}'$ is a σ-algebra of subsets of Ω'. On some measurable subset Ω of Ω' one can define conditional probability measures related to one conceptual variable given other conceptual variables, where the conceptual variable conditioned upon may or may not belong to Ω, that is, may or may not be measurable functions on $(\Omega, \mathcal{F})$, where $\mathcal{F} = \{A \cap \Omega : A \in \mathcal{F}'\}$. Strictly speaking, conditioning here must be taken as more general than the usual conditioning in statistics where we condition upon σ-algebras. We are talking in general about probabilities, given some *information*, so that we should wish to stay within the framework of propositional logic. As indicated in Appendix E, however, it seems like we need some extra assumptions in this framework to make the conditional probabilities precise in general. Therefore I will in this book stay within the probabilistic framework and limit myself to conditional probabilities given a σ-algebra as defined by (2.1). Conditional probabilities, given some conceptual variable τ which is a random variable on $(\Omega, \mathcal{F})$, is defined as the conditional probability, given the σ-algebra generated by this conceptual variable, that is, the collection of sets $\tau^{-1}(B)$, where B runs through the relevant Borel sets. Conditional probabilities, given some non-random variable τ are simply measurable functions of this variable.

When considering conditional probabilities, given the context, in most cases only part of the context will be relevant. The conceptual variables on which probabilities can be defined, will be called random variables. For simplicity, technical problems resulting from the fact that conditional distributions are only defined almost surely, are mostly disregarded in this book. However, difficulties from this in the definition of sufficiency (see Lehmann and Casella 1998) will be addressed.

A *statistical model* is defined as a conditional distribution of the data, given all parameters (together with the context, including preset values). It is assumed as usual that this class is dominated, that is, all conditional distributions are absolute continuous with respect to some fixed conditional probability measure P, given the context, where Q is defined to be absolutely continuous with respect to P if $P(A) = 0$ implies $Q(A) = 0$.

In addition, if a Bayesian analysis is to be carried out, there is a prior distribution of the parameters (again given the context). To allow for objective priors, I will, in agreement with Taraldsen and Lindqvist (2010) allow these measures to be unnormalized; see that paper and also the recent paper by McCullagh and Han (2011) on how logical difficulties with this can be avoided. Note that I talk about a Bayesian analysis to be carried out, not about Bayesian or frequentist research workers. The same person may in certain cases carry out both types of analysis, first a frequentist analysis and then at a later point of time a Bayesian analysis.

In the following, I will depart from my earlier notation and also denote random data by lower case letters. It will be clear from the context whether I talk about the pre-experimental or post-experimental situation. The statistical model will, if this is natural, be seen from a pre-experimental point of view.

As a particular case I will also look briefly at the following prediction or learning situation: In the statistical model, let y_i have some identical conditional distribution, given x_i and some fixed parameter ζ for $i = 0, 1, 2, \ldots, n$, and assume that these distributions are independent. In addition x_i $(i = 0, 1, 2, \ldots, n)$ may or may not have some identical independent distributions given a parameter κ, and ζ and κ may or may not have priors. I assume that y_0 is unknown, but the other y_i's are observed data. The x_i's are data or preset values. Thus here the UCV's—the e-variables— are y_0, ζ and κ, while the OCV's—the data z in the case where the x_i's are not preset—are $\{x_i, y_i; i = 1, \ldots, n\}$ and x_0. In principle the variables may belong to any topological space and the σ-algebras of relevance may be contained in the Borel σ-algebra, but in most practical cases they are constrained to subsets of Euclidean spaces. In this case y_0 is the e-variable of interest. This is a simple e-variable, but it is important to emphasize that it is not of the same nature as the simple e-variables of quantum mechanics; see Sect. 3.3 below.

The situation sketched above is the conceptual basis for much of Hastie et al. (2009) (supervised learning).

3.2   Data; Generalized Sufficiency and Ancillarity

Let z be a statistic, and let τ be a conceptual variable. The following two assumptions are taken as basic throughout this section and in much of this book. Any setting where these two assumptions are satisfied, will be called a *generalized experiment*.

1. The distribution of z, given τ, depends on an unknown parameter θ.

2. If τ or part of τ has a distribution, this is independent of θ. The part of τ which does not have a distribution is functionally independent of θ. However, separated from this, there may be a prior for θ which depends upon τ.

To fix ideas, think of (1) and (2) as describing a situation where inference on θ is sought from the data z in the context described by τ, but there are variants of this. In a simple experiment, z may be the whole data set, and τ may be trivial or some nuisance parameter. In addition, τ may contain the real context of the experiment, which it always will, but this is often just taken as an implicit fact. In a series of experiments, ordered in time, z may be the data set of the last experiment, and the context τ may contain some or all of the conceptual variables connected to the earlier experiments. In a meta-analysis, z may contain all data sets, and τ may contain all contexts. It is a basic condition that the model assumptions are rich enough so that (1) and (2) are meaningful. Throughout most of this section, z and τ will be held fixed.

In quantum mechanics it is not so common to distinguish between data and e-variable/parameter, because one often talks about ideal measurements. But in real measurements, such a distinction is necessary. With preparations τ, measurement data z and the parameter θ focused on, we are again in the situation of a generalized experiment; see Sect. 1.5.5. The purpose of the experiment is to gain knowledge about θ.

3.2.1 Sufficiency

We let t be a known or unknown function of z. Later I will give an example of the perhaps unfamiliar situation where we have an unknown function of the data. The concept of sufficiency was introduced by Fisher as a tool for reducing the data in a given situation without sacrificing anything related to the inference on the parameter θ. A very simple example is when we have data $z = (y_1, \ldots, y_n)$, where the y_i are assumed independent and identically $N(\mu, \sigma^2)$. Then $t(z) = (\bar{y}, s^2)$ is sufficient for $\theta = (\mu, \sigma^2)$. Here $s^2 = \sum(y_i - \bar{y})^2/(n-1)$.

Definition 3.4 We say that $t = t(z)$ is a (z, τ)-sufficient quasi-statistic for θ if the conditional distribution of z, given t, τ and θ is independent of θ. If z is the whole data set, we say just that t is τ-sufficient.

From the fact that (1) is meaningful, it follows that the conditional distribution of z, given t, τ and θ is meaningful. However, difficulties (Lehmann and Casella 1998) may arise because the conditional distribution is only defined almost everywhere. I then follow Reid (1995) in making the definition more precise: The quasistatistic $t(z)$ is (z, τ)-sufficient if there is a transformation from z to (t, v) such that the densities satisfy

$$f(z|\theta, \tau) \propto f(t|\theta, \tau) f(v|t, \tau),$$

where the constant of proportionality is independent of θ. This is a version of the factorization theorem: $t(z)$ is (z, τ)-sufficient if and only if there exist functions $g(t|\theta, \tau)$ and $h(z|\tau)$ such that for all z and θ we have

$$f(z|\theta, \tau) = g(t(z)|\theta, \tau)h(z|\tau).$$

Ordinary sufficiency results if τ is trivial, θ is the full parameter and t is a statistic. The case where part of τ is a nuisance parameter is also of interest. The general concept is of interest also in many other situations.

In general, if $t(z)$ is a (z, τ)-sufficient statistic, the rest of the distribution of z can be thought of as generated by some randomization independent of θ, and gives no information about the parameter. This will be made precise by a sufficiency principle formulated later.

It is clear that $t = z$ is a (z, τ)-sufficient statistic, but usually we are interested in smaller functions of z. In general a minimal sufficient observator will not exist, but translating a result from ordinary sufficiency theory, any boundedly complete (z, τ) observator will be minimal sufficient.

Definition 3.5 A (z, τ)-sufficient quasi-statistic t is boundedly complete if for all bounded functions h

$$E(h(t)|\theta, \tau) = 0 \text{ for all } \theta \text{ implies } P(h(t) = 0|\theta, \tau) = 1 \text{ for all } \theta.$$

Proposition 3.1 (Bahadur's Theorem) *Suppose that t takes values in a k-dimensional Euclidean space and that t is a (z, τ)-sufficient and boundedly complete quasi-statistic. Then t is a minimal (z, τ)-sufficient quasi-statistic.*

Standard results like the Rao-Blackwell Theorem and the Lehmann-Scheffé Theorem generalize immediately to (z, τ)-sufficiency. The first result says that if $g(z)$ is any estimator of a scalar θ and if $t(z)$ is a (z, τ)-sufficient statistic, then the conditional expectation of $g(z)$, given $t(z)$ is an at least as good estimator as $g(z)$, using quadratic loss. Sometimes one gets a considerable improvement using such a Rao-Blackwellization. The last result says that if t is complete and τ-sufficient for the scalar θ and $h(t)$ is an estimator of θ which is conditionally unbiased, given τ, then $h(t)$ has uniform minimal conditional variance, given τ.

Assume that we on the basis of data z want to estimate the parameter θ in the context given by τ.

Definition 3.6 If $t(z)$ is a minimal sufficient quasi-statistic for θ, and the distribution of t depends on a part of τ , we say that this part is relevant for the estimation of θ.

Example 3.8 Let $z = (y_1, \ldots, y_n)$, where $y_1, \ldots, y_n$ are independent and identically distributed (i.i.d.) $N(\mu, \sigma^2)$. Then $(\bar{y}, s^2)$ is sufficient for (μ, σ^2), where $s^2 = (n - 1)^{-1} \sum_{i=1}^{n}(y_i - \bar{y})^2$. However, even in this simple example it is of interest which parameter we focus upon. Write the log likelihood as

$$\ln f = k + \frac{1}{2\sigma^2}\sum_{i=1}^{n}(y_i-\mu)^2 + \frac{1}{2}\ln(\sigma^2) = k + \frac{1}{2\sigma^2}\sum_{i=1}^{n}[(y_i-\bar{y})^2 + n(\bar{y}-\mu)^2] + \frac{1}{2}\ln(\sigma^2).$$

From this we see:

(a) If τ contains the nuisance parameter σ^2, then $\bar{y}$ is (minimal) (z, τ)-sufficient for μ.

(b) If τ contains the nuisance parameter μ, then $\sum_{i=1}^{n}(y_i - \mu)^2$ is (minimal) (z, τ)-sufficient for σ^2. This is an example of an unknown function of the data where the concept of sufficiency is of interest.

In each case the minimality of the sufficient quasi-statistic can be proved from Proposition 3.1.

We conclude from this that σ^2 is irrelevant for the (point-)estimation of μ, while μ is relevant for the estimation of σ^2.

3.2.2   Ancillarity and Conditioning

Next I turn to the generalization of ancillarity, another basic concept introduced by Fisher.

Definition 3.7 We say that $u = u(z)$ is a (z, τ)-ancillary quasi-statistic for θ if the conditional distribution of u, given τ is independent of θ.

If u is (z, τ)-ancillary and f is a measurable function, then $f(u)$ is (z, τ)-ancillary. In the corresponding partial ordering of statistics ($u < v$ if $u = f(v)$ for some function f), z is an upper bound. By Zorn's Lemma, one or several maximal ancillaries will exist. We say that u is τ-ancillary if z is the whole data set; just ancillary if τ is trivial.

A very important and much discussed question is when one should condition upon ancillaries. Once one has conditioned upon an ancillary, this can be taken as part of the context of the experiment or the observational study. Thus the context is expanded, but after this expansion, (1) and (2) in the beginning of Sect. 3.2 will still hold. A closer discussion of the question of conditioning will be given in the next section, but as a background for this discussion we will sketch some examples.

A basic argument for conditioning is given by the following example:

Example 3.9 (Berger and Wolpert 1988; Cox 1958) Consider two potential laboratory experiments for the same unknown parameter θ such that $\mathcal{E}^1$ is planned to be carried out in New York while $\mathcal{E}^2$ is planned to be carried out in San Francisco. The owner of the material to be sent chooses to toss an unbiased coin, deciding $\mathcal{E}^1$ with probability 1/2 and $\mathcal{E}^2$ with probability 1/2. Consider the whole experiment $\mathcal{E}$

including the coin toss, and let u be the result of the coin toss. Here u is ancillary, and everybody would condition upon u in the statistical analysis.

A problem with the requirement of conditioning, is that maximal ancillaries may not be unique.

Example 3.10 Let θ be a scalar parameter between -1 and $+1$. Consider a multinomial distribution on four cells with respective probabilities $p_1 = (1+\theta)/6$, $p_2 = (2-\theta)/6$, $p_3 = (1-\theta)/6$ and $p_4 = (2+\theta)/6$ and total number of observations n. Let the corresponding observed numbers in the sample be z_1, z_2, z_3 and z_4. Recall from Sect. 2.1.2 that the multinomial distribution is a generalization of the binomial distribution with multivariate point probabilities

$$\frac{n!}{z_1! z_2! z_3! z_4!} p_1^{z_1} p_2^{z_2} p_3^{z_3} p_4^{z_4}.$$

Then one can show that each of the statistics

$$u_1 = z_1 + z_2, \quad u_2 = z_1 + z_3$$

is ancillary for θ, but they are not jointly ancillary. And conditioning upon u_1, respectively u_2 leads to distinct inference (the maximum likelihood estimator is the same, but the asymptotic variances are different).

Cox (1971) has proposed an intrinsic criterion for the choice of ancillary to condition upon in such cases, but my opinion is that this choice should depend upon the context.

Example 3.11 In a certain city the sex ratio is 1:1, and it is known that $1/3$ of the population have their own cellphone. The ratio between female and male cellphone owners is an unknown quantity $(1+\theta)/(1-\theta)$, where $-1 < \theta < 1$. One is interested in estimating θ by sampling randomly n persons from a register of the city population. It is assumed that the population is much larger than the sample size n.

Let the number of men in the sample be u_1, and let u_2 persons in the sample be owners of cellphones. Thus $u_1 = z_1 + z_2$ and $u_2 = z_1 + z_3$, where z_1 and z_2 are the male cellphone owners and non-owners, respectively, and z_3 and z_4 are the corresponding female numbers. The joint distribution of $z_1, \ldots, z_4$ is as in Example 3.10. Again each of u_1 and u_2 are ancillary, but they are not jointly ancillary.

Another question is whether or not one should *always* condition upon ancillaries. The following examples give a background for that discussion.

Example 3.12 (Helland 1995) As a part of a larger medical experiment, two independent individuals (1 and 2) have been on a certain diet for some time, and by taking samples at the beginning and at the end of that period some response like the change in blood cholesterol levels is measured. For the individual i $(i = 1, 2)$, the measured response is y_i, which is modeled as independent normal (μ_i, σ^2) with a known measurement variance σ^2.

Because the two individual have been given the same treatment (diet) in the larger experiment, the parameter of interest is not μ_1 and μ_2, but their mean: $\theta = \frac{1}{2}(\mu_1 + \mu_2)$.

Suppose now that for some reason we have only capacity to measure one of the individuals, but at the outset, we don't know which. Let u be the indicator of the individual chosen. It is clear that, given u, that is, given the person chosen, we get no information about θ. But choosing u randomly with probability $\frac{1}{2}$ for each of the two values will give us such information, provided that the identity of the individual chosen is not revealed. The last statement follows from a sampling argument: The situation is a special case of a sampling situation where n individuals are sampled randomly from a population of N individuals where the parameter of interest is the mean in the population. In addition there is a measurement error for each individual, modeled as independently normal $(0, \sigma^2)$. It is then clear that the sample mean of the observations is an appropriate estimator of this population mean. It is equally clear that this conclusion also must be valid for the special case $N = 2, n = 1$, the situation at hand.

The surprising aspect of this example is that a situation with less information can give us more ability to do inference: By not knowing u we can make some (admittedly uncertain, but nevertheless valid) inference on θ; when we know u, such an inference is impossible.

Example 3.13 Consider a sensory analysis firm where there is a staff of N trained assessors and a panel of n out of these are selected randomly to taste a particular product. A report is written. Given the assessors, what they do in the analysis must be considered as separate experiments on a common parameter θ. Consider the whole investigation, and let u be the result of choosing randomly the assessors to take part in it. Again u is ancillary. But in this case it may not be immediately natural to condition upon u in the written report.

3.2.3 Conditioning and the Conditionality Principle

Any statistical investigation has to start with a conceptual analysis. This includes choosing question of interest, collect earlier information on this question, the choice of design or sampling plan, choosing target population and sampling units, the choice of a model and maybe of a loss function etc.. The result of this analysis must be considered as a part of the context of the estimation and prediction problem. Then data are collected.

Assume now a generalized experiment, that is, a setting where assumptions (1), (2) in the beginning of Sect. 3.2 are satisfied, and let u is a (z, τ)-ancillary quasi-statistic in this setting. All the examples in Sect. 3.2.2 satisfy these assumptions.

Example 3.9 is the typical situation, and a situation where it is obvious that one should condition upon u.

To choose the conditioning in Example 3.11, one must specify further. Suppose that the data collection is done by first finding out whether the person in question is a man or woman, thereafter asking about cellphone ownership, then the conditioning should be done upon sex. In the opposite case, if the data collection is done from a register of cellphone owners, later asking about sex, then one should condition upon cellphone ownership. In the case where the data are found from a register containing both information on sex and cellphone ownership, one should perhaps condition upon both variables, even though we don't have joint ancillarity here.

The last two examples describe different situations. First consider Example 3.12. Here the parameter of interest θ is a function of the inaccessible conceptual variable $\phi = (\mu_1, \mu_2)$, and an ancillary for θ is the choice u of a person to investigate. This is chosen randomly, and is at the outset unknown to the experimentalist. Assuming that u takes some definite value, let μ_u be the value of μ for the specific person chosen. The point is that information about u may give information about μ_u, but no information about the parameter of interest $\theta = \frac{1}{2}(\mu_1 + \mu_2)$.

The Generalized Principle for Conditioning (GPC) *Assume that u is a maximal (z, τ)-ancillary quasi-statistic for an parameter θ.*

1. *Assume that the generalized experiment obtained by knowing the value of u has θ as the parameter. Then one should condition upon u. If there are several maximal such u's to choose between, one should condition upon the one corresponding to the data that have first been obtained.*
2. *If knowledge of u will imply a generalized experiment with another parameter than θ, one should not condition upon u.*

Part (1) is consistent with the conditioning chosen in Example 3.9 and in Example 3.11. This is the most common situation, and is often thought about as just the principle of conditioning. Part (2) is consistent with the decision not to condition in Example 3.12, since the parameter in the experiment obtained by conditioning is μ_u, not θ, which is the parameter of interest.

Example 3.13 is a bordering situation. In most cases, a user of the results of the sensory analysis will not be interested in which assessors that are chosen, will not ask for this information and will thus not condition upon this information. More precisely, if one conditions, the parameter of the conditioned experiment will be the mean for the n chosen assessors, while the parameter of interest will most probably be a population mean.

I have given a normative form of the conditionality principle. For the further development in this book it is also important to consider a descriptive form, which is often given in the literature; see Berger and Wolpert (1988). In this case, the notion 'one should condition upon …' translates into '…the unconditioned experiment contains no experimental evidence on θ in addition to that of the conditioned experiment'. As in Berger and Wolpert (1988), the concept of 'experimental evidence' is left undefined, i.e., it can be made precise in any reasonable way.

The Generalized Weak Conditionality Principle (GWCP) Suppose that there are two generalized experiments E_1 and E_2 with common parameter θ and with equivalent contexts τ. Consider the mixed experiment E^*, whereby $u = 1$ or 2 is observed, each having probability $1/2$ (independent of θ, the data of the experiments and the contexts), and the experiment E_u is then performed. Then the evidence about θ from E^* is just the same as the evidence from the experiment actually performed.

Note that this corresponds to the situation 1) of the GPC: The variable u is a statistic here; the two experiments are known to the experimentalist. Two contexts τ and τ' are defined to be equivalent if there is a one-to-one correspondence between the relevant parts of them: $\tau' = f(\tau), \tau = f^{-1}(\tau')$.

3.2.4 The Sufficiency and Likelihood Principles

The motivation behind the definition of a sufficient statistic is that one wants to reduce the data and still get the same information about the parameter. One version of the sufficiency principle, as formulated in Berger and Wolpert (1988), translates to our setting as follows:

The Generalized Weak Sufficiency Principle (GWSP) Consider a generalized experiment in a context τ as described above, let z be the data of that experiment, and let θ be the parameter connected to the experiment. Assume that (1) and (2) of Sect. 3.2 are satisfied. Let $t = t(z)$ be a (z, τ)-sufficient statistic for θ. Then, if $t(z_1) = t(z_2)$, the data z_1 and z_2 contain the same experimental evidence about θ in the context τ.

There can be given many examples to support the GWSP. The simplest example is an independent measurement series $z = (y_1, \ldots, y_n)$, where the y_i's are normal (μ, σ^2). If σ^2 is known, $\bar{y} = n^{-1} \sum y_i$ is sufficient for μ, and any reasonable inference is based upon $\bar{y}$. If σ^2 is unknown, then $t(z) = (\bar{y}, s^2)$ is sufficient for $\theta = (\mu, \sigma^2)$, where $s^2 = (n-1)^{-1} \sum (y_i - \bar{y})^2$. (The denominator $n-1$ makes s^2 an unbiased estimator of σ^2.) Any reasonable inference on θ under the normal model is based upon $t(z)$. This kind of data reduction was Fisher's motivation for introducing the concept of sufficiency.

The general argument for the GWSP runs as follows: By Fisher's factorization theorem, if $t(z)$ is sufficient for θ, then the probability density can be written as $f(z|\theta) = g(t(z)|\theta)h|z|$, where $h(z)$ is independent of θ. But then an artificial experiment, where $h(z)$ is generated by a computer simulation, will give the same information an θ as the real experiment.

Now following an argument from Berger and Wolpert (1988), using the GWSP and the GWCP, which we will regard as more or less obvious, we can derive the following likelihood principle. This result is a classical theorem first given by Birnbaum (1962). The argument is reproduced for completeness in Appendix A for the discrete case; this is in fact the case I need later in the discussion of quantum mechanics. For the continuous case, see Berger and Wolpert (1988).

A version of the likelihood principle will be used later to motivate Born's formula in quantum mechanics.

The Generalized Likelihood Principle *Consider two generalized experiments with equivalent contexts τ, and assume that θ is the same full parameter in both experiments. Suppose that two observations z_1^* and z_2^* have proportional likelihoods in the two experiments, where the proportionality constant c is independent of θ. Then these two observations produce the same experimental evidence on θ in this context.*

Since both my definition of ancillary and my definition of sufficient statistic depend on the context, and therefore the context is kept fixed in the corresponding principles, it is important that it is kept essentially fixed also here. This aspect makes the generalized likelihood principle weaker than the principle as formulated in the literature, in particular in Berger and Wolpert (1988). On the other hand, paradoxes like what the ordinary likelihood principle seems to imply in the following situation are avoided.

Example 3.14 Suppose that $s_1, s_2, \ldots$ are independent, identically distributed variables with $P(s = 1) = \theta$ and $P(s = 0) = 1 - \theta$, i.e., iid Bernoulli variables with parameter θ. In experiment $\mathcal{E}_1$, a fixed sample size of ten observations is decided upon, and the sufficient statistic $t_1 = \sum_{i=1}^{10} s_i$ turns out to be $t_1 = 8$. In experiment $\mathcal{E}_2$, it is decided to take observations until a total of 2 zeroes has been observed. Then assume that the sufficient statistics $t_2 = \sum s_i$ also turns out to take the value 8. The two likelihoods are proportional, but the contexts are different, so the intuition that the two experiments may lead to different inference on θ is supported by my version of the likelihood principle. For further discussion of this example, see Berger and Wolpert (1988) and references there.

The introduction of a context makes my formulation of the likelihood principle far less controversial than the ordinary formulation. According to the ordinary principle, the way data are obtained is irrelevant to inference; all information is contained in the likelihood. Thus sampling plans, randomization procedures, and stopping rules are irrelevant according to a common interpretation of the ordinary principle. Furthermore, common frequentist concepts like bias, confidence coefficients, levels and powers of statistical tests, etc., are irrelevant, as they depend on the sample space, not only on the observed observations. In my formulation, all these concepts are related to the context. Also Bayesian priors, if needed, are contained in the context. Maximum likelihood estimation can not be derived from the likelihood principle, but is obviously permissible as a method of obtaining reasonable proposals for estimates in general.

An important special case of the generalized likelihood principle is when the proportionality constant c is equal to 1. Then the two observations z_1^* and z_2^* have equal likelihoods. Again an important special case is when the two experiments are identical. A consequence of the generalized likelihood principle is then that all experimental evidence, given the context, is a function of the likelihood of the experiment, i.e, is contained in the likelihood function.

3.3 Prediction and Simple e-Variables

Now look at situations with what I called in Chap. 1 simple e-variables. In most statistical settings this corresponds to the area of prediction. So first, let y is an unknown e-variable (random variable) to be predicted, we have data z and a statistical model for (z, y) depending on an unknown parameter η. The context is denoted by τ. There are many situations where this is a relevant practical problem; one situation is sketched at the end of Sect. 3.1.

Berger and Wolpert (1988) formulated a likelihood principle for prediction, stating that *all evidence about (y, η) is contained in the likelihood function*

$$L_z(y, \eta|\tau) = p_\eta(y, z|\tau). \tag{3.1}$$

In Bjørnstad (1990) predictive likelihood is reviewed, taking this as a point of departure. The essential problem is that of eliminating η from (3.1) using one of the following operations on L_z: integration, maximization and conditioning. In all, 14 different combinations of these operations were proposed, resulting in 14 different versions of the what is called a predictive likelihood. In a statistical setting, it is too much to demand that all evidence about θ is contained in such a predictive likelihood.

The situation is completely different in quantum mechanics. Here, in a given situation, we will have a model for the data z depending upon the context τ and the e-variable of interest θ; see the example in Sect. 1.5.5. Thus, even though θ is simple, from a statistical point of view it acts as a parameter in the model. An eventual extra parameter η in such a model will be assumed known from earlier experiments of the same type, and may be included in the context. This gives a unique likelihood $L(\theta|z, \tau) = p(z|\tau, \theta)$. And in this situation the whole discussion in Sect. 3.2 is valid. In particular the generalized likelihood principle above holds true.

In simple quantum theory θ will be discrete. The situation will be taken up again in Sect. 5.4, where the main point is to focus on a particular θ.

3.4 Epistemic Processes, Decisions and Actions

Recall that an epistemic process may result in a confidence interval or a credibility interval for the parameter θ. Also recall that (cp. Sect. 2.1.1) in the present book we support all 3 interpretations of the probability concept: The one based on symmetry, the subjective interpretation and the frequentist interpretation. The subjective interpretation is relevant if the context contains a prior. The frequentist interpretation is relevant if the epistemic process can be repeated an arbitrary number of times with the same context. The first case gives a credibility interval; the second case a confidence interval. An important case for us is when we have a prior which is a right invariant measure of a transitive group, and the same group acts

on the sample space. Then (Helland 2010, Corollary 3.6.2) the confidence intervals
and the credibility intervals with the same coefficient are numerically equal if they
are based on equivariant estimators (estimators which transform in the same way as
the corresponding parameters under the group).

Assume now that the epistemic process results in a 99% confidence interval or
credibility interval $[\underline{\theta}, \bar{\theta}]$ for the parameter θ. Then the next step is an existential
choice, a decision: We claim that in the given context θ really belongs to the
interval $[\underline{\theta}, \bar{\theta}]$. In this way a decision is always a part of the epistemic process.
This decision may be the basis for further decisions, for instance in the planning of
new experiments.

A special case is when θ is discrete. Then the interval may degenerate into a
single point $\underline{\theta}$. The epistemic decision may then be that θ really takes the value $\underline{\theta}$.
Similar kinds of decisions will be important in the second part of the book.

For those who know about hypothesis testing: A Neyman-Pearson hypothesis
testing situation may be taken as mathematically equivalent to a one-sided or two-
sided confidence interval situation. Suppose that we want to test the null-hypothesis
$H_0 : \theta = \theta_0$. Then the nullhypothesis is rejected with the level α if and only if θ_0 do
not belong to a confidence interval with coefficient $1 - \alpha$.

Another decision type is a Fisherian p-value testing. In this case a nullhypothesis
and a direction of rejection are chosen as a part of the context. The result of the
decision is a p-value, 1 minus the confidence coefficient of the largest one-sided
confidence interval not containing θ_0.

Both in science and in everyday life the epistemic process may be a part of
a larger process where a goal is formulated. The epistemic process itself may be
formal or informal. A great variety of decisions to act may result after an epistemic
process is carried out, and the result of this process is known.

A scientist who has gained some knowledge through an empirical process may
or may not decide to publish the results. The publishing process will involve several
minor decisions. In a similar way, any human being that has gained some knowledge
may decide to share his knowledge with other persons. But his newly gained
knowledge may also result in a wide variety of other decisions.

Every serious communication between human beings involves a complicated
interplay between epistemic processes and decisions together with other elements
like insights, reflections and appreciations. In this interplay, the free will of each
individual involved plays an important role. Intuition is also important.

A group of persons may decide to go collectively into an epistemic process, and
to make collective decisions on how to act after the epistemic process is completed.

Each decision on how to act should be taken in an intelligent way, taking into
account the knowledge available. The time sequence will be important: First an
epistemic process, then a decision or a set of decisions, then actions and after that
perhaps new epistemic processes.

A person or a group of persons may decide explicitly to go into an epistemic
process before making a new important decision. The result of the epistemic process
will then be a part of the context of the important decision.

In general the result of an epistemic process may form part of the context for new decisions or for other epistemic processes. The time span here may be short for some decisions or epistemic processes; often longer for more important decisions or epistemic processes.

In scientific decisions it is always necessary to focus. Such focusing should be done at every stage of the decision process. The focusing made when choosing the epistemic question will play an important role in the second part of this book.

In the next part of the book I will address quantum mechanics from an epistemic point of view. A crucial concept is then that of an inaccessible conceptual variable, that is, a conceptual variable for tentative use in an epistemic processes which cannot be estimated with arbitrary accuracy by any experiment. In the regression model where the dimension p by necessity is larger than the number n of observations, the regression vector β must be seen as an inaccessible conceptual variable. However, under suitable circumstances, the e-variable function $\beta^T x_0$ may still be estimable. In particular this is the case when x_0 is regarded as random, with the same distribution as the other x_i's. One approach towards estimating this function may be model reduction as discussed earlier.

References

Berger, J. O., & Wolpert, R. L. (1988). *The likelihood principle.* Hayward, CA: Institute of Mathematical Statistics.

Birnbaum, A. (1962). On the foundation of statistical inference. *Journal of the American Statistical Association, 57*, 269–326.

Bjørnstad, J. F. (1990) Predictive likelihood: A review. *Statistical Science, 5*, 242–265.

Cook, R. D. (2007). Fisher lecture: Dimension reduction in regression. *Statistical Science, 22*, 1–26.

Cox, D. R. (1958). Some problems connected with statistical inference. *Annals of Statistics, 29*, 357–372.

Cox, D. R. (1971). The choice between ancillary statistics. *Journal of the Royal Statistical Society. Series B, 33*, 251–255.

Fisher, R. A. (1922). On the mathematical foundations of theoretical statistics. *Philosophical Transactions of the Royal Society of London. Series A* **222**, 309-368. Reprinted in: Fisher R. A. Contribution to Mathematical Statistics. Wiley, New York (1950)

Hastie, T., Tibshirani, R., & Friedman, J. (2009). *The elements of statistical learning. Data mining, inference, and prediction.* Springer series in statistics. Berlin: Springer.

Helland, I. S. (1995). Simple counterexamples against the conditionality principle. *The American Statistician, 49*, 351–356. Discussion *50*, 382–386.

Helland, I. S. (2010). *Steps towards a unified basis for scientific models and methods.* Singapore: World Scientific.

Lehmann, E. L., & Casella, G. (1998). *Theory of point estimation.* New York: Springer.

McCullagh, P., & Han, H. (2011). On Bayes's theorem for improper mixtures. *Annals of Statistics, 39*, 2007–2020.

Reid, N. (1995). The roles of conditioning in inference. *Statistical Science, 10*(2), 138–157.

Stigler, S. M. (1976). Discussion of "On rereading R.A. Fisher" by L.J. Savage. *Annals of Statistics, 4*, 498–500.

Taraldsen, G., & Lindqvist, B. H. (2010). Improper priors are not improper. *The American Statistician, 64*, 154–158.

Chapter 4
Towards Quantum Theory

Abstract The point of departure is that of an inaccessible conceptual variable; several examples are given, both macroscopic and microscopic. One focuses on a certain e-variable, a function the inaccessible conceptual variable. Some questions on the foundation of quantum theory are briefly addressed. Next, a maximal symmetric setting is defined, where certain symmetry assumptions are introduced. Spekkens' toy model is discussed and is related to this setting. A concrete Hilbert space is defined, and it is shown that under a certain technical condition question-and answer pairs concerning e-variables are in one-to-one correspondence with unit vectors of this Hilbert space. Finally, a more general symmetric setting, corresponding to degenerate eigenvalues of the relevant operator, is discussed.

4.1 Inaccessible Conceptual Variables and Quantum Theory

The statistical literature is full of discussions on how to do inference, but contains very little on the choice of question to do inference on in some given situation. These different questions may be conflicting, even complementary. In the following sections I will start by formalizing a way in which the discussion of such complementary questions may be addressed in the extreme case where it is only possible to raise one out of many different possible questions at a time. Each such question will be an epistemic question 'What is θ?' for some e-variable θ, and I will assume that the epistemic process ends by giving some information about θ, in the simplest case a complete specification: $\theta = u_k$.

The concept of an epistemic process is taken to be very wide in this book. In addition to statistical questions concerning a parameter θ, we can think of questions like: How many sun hours will there be here tomorrow? At the outset, to address this epistemic question will involve meteorological expertise and a lot of data from similar situations, but tomorrow the question can be answered by just counting the number of sun hours. Both these processes will be seen as epistemic processes.

However, when it comes to the parameters/e-variables of the epistemic processes, I will often make more specific assumptions. I will then take generalized experiments as point of departure, that is, I assume that there in each setting exist data z and a context τ such that the assumptions (1) and (2) of Sect. 3.2 are satisfied. For

© Springer-Verlag GmbH Germany, part of Springer Nature 2018

I. S. Helland, *Epistemic Processes*, https://doi.org/10.1007/978-3-319-95068-6_4

the most part, z and τ will be implicit in the discussion, but they will be there. The e-variables are also assumed to be associated with some actor (observer) or with a group of communicating actors.

So far I have assumed that each conceptual variable relevant to an epistemic process is accessible, that is, it can be estimated or given a value with arbitrary accuracy by any experiment. In Helland (2006, 2008, 2010) several situations with inaccessible conceptual variables were described (see also below), and it was indicated that such situations in special cases could form a link to important parts of quantum theory. I consider this way of thinking to be essential as a step towards obtaining a unification of epistemic science, and also as an attempt to give an alternative background for the—from a statistical point of view and also from the layman's point of view—very formal language that one finds in textbooks and in scientific publications, both within quantum physics and in the mathematical traditions developed from this. In the following sections a less formal approach will be presented. Compared to my earlier publications, the discussion here will hopefully give both a simpler and a more complete treatment of my approach towards quantum mechanics.

In statistics, the parameter concept is connected to a hypothetical population of items. My e-variables are intended also for situations where we have a single item or a few items, and a human subject or a group of subjects use these variables in making statements about the item(s). This is crucial for my epistemic interpretation of quantum mechanics, an interpretation which I also share with the Bayesian quantum foundation school (QBism); see below.

The concept of e-variable will be very important in this book. Recall that it is any conceptual variable used in the epistemic process. In the same way as a parameter in statistics is any property for an hypothetical population, an e-variable can in principle be any property of a population, a single unit or a group of units. Like a parameter, it is a theoretical variable before the epistemic process, but after the process the observer or the set of communicating observers in question have some information about the value of the e-variable. An e-variable is a property of a unit or a set of units. A modern view of quantum theory and particle physics, see Kuhlmann (2013), reduces everything to properties and relations.

Quantum theory has a long history starting with the work of several eminent physicists in the beginning of the previous century, via the formalization made by von Neumann (1932) to the rather intense debate on quantum foundation that we see today. Several good books on quantum theory exist, for instance Ballentine (1998). Interpretations of the theory have been given by many authors, but it has also been argued that no interpretation is needed; see Fuchs and Peres (2000). Several authors have derived quantum theory from a few explicit or implicit physical assumptions; see Hardy (2001), Chiribella et al. (2010), Masanes (2010), Fields (2011), Fivel (2012) and Casinelli and Lahti (2016). There is also a group of quantum foundation researchers working towards a link with Bayesian inference; see Caves et al. (2002), Schack (2006), Timpson (2008), Fuchs (2010), Fuchs and Schack (2011) and Fuchs et al. (2013). The use of quantum information theory in the exploration of the foundation has also recently proved to be very useful, see Fuchs (2002). The present work has much in common with these schools, but I find it fruitful to maintain a broader link to statistics, in particular to allow a broader view on statistical inference than just the Bayesian view. In this way I will argue for a foundation which is purely epistemological: A general approach for going from experienced data to information about the nature behind these data. I will discuss connections to the quantum Bayesian interpretation later; see also Chap. 1.

One very obvious case of an inaccessible conceptual variable is in connection to counterfactual reasoning. Assume a single medical patient and let the doctors have the choice between two mutually exclusive treatments. Let θ^i be the time for this patient until recovery when treatment i is used ($i = 1, 2$), and let $\phi = (\theta^1, \theta^2)$. Then θ^1 or θ^2 can be predicted before the treatment is applied, and each of them can be determined precisely after some time period, but ϕ is inaccessible, that is, there is no procedure by which ϕ can be given a value with arbitrary accuracy at any time for a single patient by any medical doctor, by any scientist or by any observer. This can be amended by instead of one patient considering large homogeneous groups of patients, which is done in standard statistical texts, but in practice there is a limitation on how homogeneous a group of patients can be. And concepts may be of interest for one single patient, too.

Here are two other examples of inaccessible conceptual variables:

- We want to measure some quantity θ^1 with a very accurate apparatus which is so fragile that it is destroyed after a single measurement. There is another quantity θ^2 which can only be found by dismantling the apparatus, and then it can not be repaired. The vector $\phi = (\theta^1, \theta^2)$ is again inaccessible.
- Assume that two questions are to be asked to a single individual at some given moment, and that we know that the answer will depend on the order in which the questions are posed. Let the e-variable (θ^1, θ^2) be the answers when the questions are posed in one order, and let the answers be (θ^3, θ^4) when the questions are posed in the opposite order. Then the vector $\phi = (\theta^1, \theta^2, \theta^3, \theta^4)$ is inaccessible.

Now go to the quantum mechanical situation. It is well known that the position θ^1 of a particle can be measured accurately in some experiments and its momentum θ^2 can be measured accurately in other experiments, but that the vector $\phi = (\theta^1, \theta^2)$ is inaccessible. Similarly, the spin vector ϕ of a particle is inaccessible, but its component θ^a in some fixed, determined direction a is possible to measure in a suitable experiment.

In general, let $\phi = (\theta^1, \theta^2)$ be inaccessible. Then different experimental settings are needed to measure θ^1 and θ^2. In the words of Niels Bohr, the variables θ^1 and θ^2, which I call e-variables, are complementary. The concept of complementarity was crucial to Bohr, and it has been crucial to the foundation of quantum mechanics even before its formal apparatus was developed. In the same way, the concept of an inaccessible conceptual variable will be crucial in the further development of this book.

From a statistical point of view: Inaccessible parameters also occur in linear models of non-full rank, often used in the case of unbalanced data, cp. Searle

(1971), and in the analysis of designed experiments where only some contrasts can be estimated. Also, in regression models where the number of variables by necessity is larger than the number of observations, the regression parameter is an inaccessible parameter. In my opinion a more complete theory of statistical inference is definitely obtained if we allow for inaccessible conceptual variables.

It is a crucial fact that the inaccessible conceptual variables ϕ are abstract variables in some mathematical space and that operations such as group actions may be made on this space. This is the case with the counterfactual example above, where a group action such as a change of time scale can be made. See the summary of group theory in Appendix B. In general, let ϕ vary in a set Φ. Then the group of endomorphisms on Φ is the group of all possible transformations of elements of Φ. This group always exists from a mathematical point of view. In my later approach towards quantum mechanics, I will choose a fixed subgroup G of the group of endomorphisms acting on the space Φ of inaccessible conceptual variables. Important subgroups of G again, are the groups G^a, where G^a corresponds to all transformations of the values of $\theta^a = \theta^a(\phi)$ in the space Θ^a.

What is important to note, however, is that I will not regard the inaccessible conceptual variables as physical variables, and they do not take concrete values, so I am not developing a hidden variable theory of the kind that has been much debated in the physical literature over the years. Also, the e-variables/ parameters are not hidden variables, but closely connected to the epistemic process. Note that the parameters of statistics exist only in our minds.

Historically, an example of a hidden variable theory is David Bohm's dual wave-particle theory, and John Bell (see Bell 1987) proved that this theory is non-local. In fact, Bell proved much more. His famous theorem states that any realistic theory consistent with quantum mechanics must be non-local. This result has been very important in discussions among physicists in recent years. Bell's theorem is proved using what is called the Einstein-Podolski-Rosen experiment and Bell's inequality, concepts which will be discussed later in this book. One point for me here is that I do not want to develop a non-local theory, that is, a theory where communication is made by signals traveling faster than the light speed. Then I am instead forced to take a closer look upon the concept of realism. This has also been done recently in a very convincing way by Nisticò and Sestito (2011). In that paper they take as a point of departure the criterion of reality as formulated be Einstein et al. (1935):

Criterion of Reality *If, without in any way disturbing a system, we can predict the value of a physical quantity, then there exists an element of physical reality corresponding to this physical quantity.*

Following arguments from Bohr's discussion of Einstein et al. (1935) they make the case for a strict interpretation of this criterion:

Strict Interpretation *To ascribe reality to P, the measurement of an observable whose outcome would allow for the prediction of P, must actually be performed.*

Nisticò and Sestito (2011) go on and formulate an extension of quantum correlation which is consistent with the strict interpretation, and using this they show that Bell's argument and several related arguments in the literature fail when realism is interpreted in this strict way. Thus the possibility turns out to be open to interpret the non-locality theorems in the physical literature as arguments supporting the strict criterion of reality, rather than as a violation of locality.

Since the present book is theoretical and not experimental, I will have to modify Nisticò and Sestito's requirement of strict interpretation slightly: '…a description of how the measurement can be actually performed, must be given.' It is important that my conceptual variables are thought of as defined by one person or a group of persons and to the experimental data that he/she/they are able to obtain.

In other papers, Bell's theorem is interpreted as saying that quantum physics must necessarily violate either the principle of locality or counterfactual definiteness. Counterfactual definiteness is defined as the ability to speak with meaning of definiteness of results of measurements that have not been performed (i.e., the ability to assure the existence of objects, and properties of objects, even when they have not been measured.) In this book it is crucial that I do not assume counterfactual definiteness. All my conceptual variables are assumed to be defined by some person(s), and these conceptual variables will not necessarily be such that results of measurements not performed will have meaning. Here is a simple example: By first sight, one of the statements 'I have something on my lap' and 'I do not have anything on my lap' must be true. But if I am standing, neither of these statements are true. The logical status of statements must depend on the context.

In my formulation, I will look upon the accessible e-variables as variables connected with experiment which actually can be imagined to be performed by some person. This person will have a certain context for his experiment. It is possible that another person, who has no communication with the first one, has a different context and uses different e-variables to formulate his observations, therefore getting seemingly conflicting predictions. But as soon as communication is restored, there must be no conflict any more. To make this precise: The two persons must then make non-conflicting predictions if they agree on a common context, and they must agree on observed results as long as they both have observed results.

In this chapter I will assume ideal experiments, where I will not distinguish between data and corresponding e-variables. I will come back to more realistic experiments with data in the next chapter.

4.2   The Maximal Symmetrical Epistemic Setting

I proceed to discuss a setting from which I will show that essential parts of the formalism of quantum mechanics can be derived under certain technical conditions. From my point of view this is nothing but a special situation with an inaccessible conceptual variable, where I focus upon accessible sub conceptual variables and where symmetry is introduced by natural group actions. The purpose at this particular point is not to derive all aspects of quantum mechanics, only as much that we see that the e-variable concept is useful also in this connection, so that we can obtain an interpretation where there is a link to the ordinary statistical theory of estimation. The rest of this chapter will involve some technical discussions, and can only be skimmed in the first reading of the book. The results of these discussions are crucial, however: To which extent can the Hilbert space formalism be derived from simple assumptions in the epistemic process setting? The next chapter will begin by simply assuming this formalism.

Let in general ϕ be an inaccessible conceptual variable taking values in some topological space Φ, and let $\lambda^a = \lambda^a(\phi)$ be accessible functions for a belonging to some index set $\mathcal{A}$. I will repeat that a conceptual variable is accessible if it in the given context can be estimated with arbitrary accuracy by *some* experiment. In other words, the λ^a's are e-variables. Technically I will without further mention assume that all functions defined on Φ are Borel-measurable. To begin with, I will assume that the functions λ^a are maximal, and also that there is an one-to-one functional relation between them. This is made precise below. In general, transformations of Φ by group elements g may be defined.

Assumption 4.1

a) *Consider the partial ordering defined by $\alpha \leq \beta$ iff $\alpha = f(\beta)$ for some function f. Under this partial ordering each $\lambda^a(\phi)$ is maximally accessible, that is, (1) $\lambda^a(\phi)$ is accessible, an e-variable; (2) if $\lambda^a(\phi) = f(\mu(\phi))$ for a non-invertible function f, then $\mu(\phi)$ is inaccessible.*

b) *For $a \neq b$ there is an invertible transformation g_{ab} such that $\lambda^b(\phi) = \lambda^a(g_{ab}(\phi))$.*

Note that the partial ordering in a) is consistent with accessibility: If β is accessible and $\alpha = f(\beta)$, then α is accessible. Also, ϕ is an upper bound under this partial ordering. The existence of maximal accessible conceptual variables follows then from Zorn's lemma.

To be clear, no summation convention is used in b). This assumption induces an one-to-one functional relation between λ^a and λ^b.

In this abstract setting, the inaccessible variable ϕ and the accessible variables $\lambda^a(\phi)$ can be anything. However, to begin with it might be useful to have the following physical example in mind: Let ϕ be the spin or angular momentum vector for a particle, then focus upon some direction a in space, and let $\lambda^a(\phi)$ be the spin or angular momentum component in that direction. Let the group elements g consist of rotations of ϕ in space. It is useful to think through what Assumption 4.1 means in this setting. An important point of departure is that ϕ only exists in our minds. A closer discussion of this example will be given below.

Consider in general the situation where the vector $\phi = (\lambda^1, \lambda^2)$ is inaccessible. Then the statement that λ^1 and λ^2 are maximally accessible is equivalent to the statement that they are complementary in Niels Bohr's sense. The concept of complementarity is extremely important in quantum mechanics. In Sect. 6.3 I will discuss the concept in other contexts as well.

Below, I will often single out a particular index $0 \in \mathcal{A}$. Then (a), given (b), can be formally weakened to the assumption that $\lambda^0(\phi)$ is maximally accessible, and b) can be weakened to the existence for all a of an invertible transformation g_{0a} such that $\lambda^a(\phi) = \lambda^0(g_{0a}(\phi))$. Take $g_{ab} = g_{0a}^{-1} g_{0b}$.

In the example above with counterfactual medical treatments, we can take $\lambda^a = \theta^1$, $\lambda^b = \theta^2$, $\phi = (\lambda^a, \lambda^b)$ and $g_{ab}((\lambda^a, \lambda^b)) = (\lambda^b, \lambda^a)$. In general, when the transformation of Assumption 4.1b) exists, it is usually easy to see how it can be chosen.

Even though ϕ is inaccessible, it is possible to operate on ϕ with functions, in particular group actions. The group of endomorphisms on Φ, all transformations of elements ϕ always exists from a mathematical point of view, and one can imagine many subgroups of this group. Some of these will now be defined.

Definition 4.8 For each a, let $\tilde{G}^a$ be the group of endomorphisms on Λ^a, the space upon which λ^a varies. For $\tilde{g}^a \in \tilde{G}^a$ let g^a be any transformation on Φ for which $\tilde{g}^a \lambda^a(\phi) = \lambda^a(g^a \phi)$.

Note that this makes sense in the spin/ angular momentum case: For some fixed integer or half integer number j, the possible values of $\lambda^a(\phi)$ are $-j, -j + 1, \ldots, j - 1, j$. Start with some vector ϕ, and fix a plane through this vector. Then by suitable rotations in this plane, $\lambda^a(\phi)$ will change from one of these values to another arbitrary value. $\lambda^a(\phi)$ is fixed when this plane is rotated with fixed ϕ.

It is easily verified in general that

1. For fixed $\tilde{g}^a$ the transformations g^a form a group.
2. For fixed a the transformations g^a form a group G^a.

In simpler terms, the group G^a is the group transforming values of λ^a into the same or other values of this e-variable, and the corresponding group $\tilde{G}^a$ is the group of all transformations of these values.

The group G^a can be characterized as follows. Let a function η on Φ be called permissible (Helland 2010) with respect to a group H if $\eta(\phi_1) = \eta(\phi_2)$ implies $\eta(h\phi_1) = \eta(h\phi_2)$ for all $h \in H$. Then G^a is the maximal group under which λ^a is permissible.

Obvious consequences of Definition 4.8 are that $\tilde{G}^a$ is transitive over Λ^a and that G^a is transitive over Φ.

Now single out a fixed index $0 \in \mathcal{A}$.

Definition 4.9 Let G be the group of transformations generated by G^0 and the transformations g_{0a}, $a \in \mathcal{A}$.

It is easily verified that $G^a = g_{0a}^{-1} G^0 g_{0a}$. Together with $g_{ab} = g_{0a}^{-1} g_{0b}$ this implies that G also is the group generated by G^a, $a \in \mathcal{A}$ and g_{ab}, $a, b \in \mathcal{A}$.

Now I want to introduce the further

Assumption 4.2

a) *The group G is a locally compact topological group, and satisfies weak assumptions such that an invariant measure on Φ exists. (see Appendix B).*

b) *The group generated by products of elements of $G^a, G^b, \ldots; a, b, \ldots \in \mathcal{A}$ is equal to G.*

Assumption 4.2a) is a technical one, needed in the next section. Note that G is defined in terms of transformations upon Φ, so that the topology must be introduced in terms of these transformations. Technically this can be achieved by assuming Φ to be a metric space with metric d, and letting $g_n \to g$ for instance if $\sup_\phi d(g_n(\phi), g(\phi)) \to 0$. Concerning Assumption 4.2b), it follows from $g^a g^b \ldots = g_{0a}^{-1} g^0 g_{0a} g_{0'b}^{-1} g^{0'} g_{0'b}, \ldots$, where $g^a \in G^a, g^b \in G^b, \ldots$ and $g^0, g^{0'}, \ldots \in G^0$, that the group of products is contained in G. That it is equal to G, is an assumption on the richness of the index set $\mathcal{A}$ or the richness of G^0.

The setting described here, where Assumptions 4.1 and 4.2 are satisfied, includes many quantum mechanical situations including spins and systems of spins. I will call it *the maximal symmetrical epistemic setting*. Later I will also sketch a macroscopical situation where the assumptions of the maximal symmetrical epistemic setting are satisfied. However, the focus in the present book will be quantum-mechanical.

An important special case is when each Λ^a is discrete. Then $\tilde{G}^a$ is the group of permutations of elements of Λ^a, and G^a is the group of all transformations on Φ which induce permutation of Λ^a. In this situation I will later define a state of the system as a focused question: "What is the value of λ^a?" together with a definite answer: "$\lambda^a = u_k$". Under an additional technical assumption on the group structure, I will prove that this leads to a link to the ordinary Hilbert space formalism of quantum mechanics.

Example 4.15 Model the spin vector of a particle such as the electron by a vector ϕ, an inaccessible conceptual variable. More generally, we can let ϕ denote the total spin/angular momentum vector for any particle or system of particles. Let Φ be the sphere corresponding to a fixed norm $\|\phi\|$. Let G be the group of rotations of this vector.

Next, choose a direction a in space, and focus upon the spin component in this direction:

$$\zeta^a = \|\phi\|\cos(\phi, a).$$

Associate $\zeta^a(\phi)$ with the group F^a of rigid rotations around a together with a reflection in a plane through the origin perpendicular to a. This is the largest subgroup of G with respect to which ζ^a is a permissible subvariable. (For a closer discussion of the concept of permissible subparameter; see Helland 2010, Chapter 3.) The actions of the group $\tilde{F}^a$ on ζ^a are just a change of sign together with the identity.

Finally, introduce model reduction of the kind discussed in Sect. 2.2: The orbits of $\tilde{F}^a$ as acting on ζ^a are given as two-point sets $\{\pm c\}$ together with the single point 0. A maximal model reduction is to one such orbit. Later I will give arguments to the effect that we want to reduce to the a set of orbits indexed by an integer or half-integer j, and that we will let this reduced set of orbits be

$$-j, -j+1, \ldots, j-1, j,$$

this together with $\|\phi\|^2 = j(j+1)$.

Now fix j and let λ^a be the conceptual variable ζ^a reduced to this set of orbits of $\tilde{F}^a$. This is assumed to be the maximal accessible e-variable. Define the transformations g_{ab}, and define the groups $\tilde{G}^a$, G^a and G as in the maximal epistemic setting. It is easy to see that the group G is as before. The group $\tilde{G}^a$ is the group of permutations of values of λ^a. The group elements g^a can be seen as products of two kinds of elements. The first kinds are rotations around a. The second kinds are suitable rotations of each ϕ in a plane through a and ϕ.

We can prove that the general assumptions of this section are satisfied. In the case $j = 0$, where we must define $G = G^0$ to be the trivial group. Otherwise G is the group of all rotations of vectors ϕ, is obviously compact and satisfying Assumption 4.2a). Here an argument leading to the proof of Assumption 4.2b): Given a and b, a transformation g_{ab} sending $\lambda_a(\phi)$ onto $\lambda_b(\phi)$ can be obtained by a reflection in a plane P perpendicular to a plane containing the two vectors a and b, where P contains the mid-line between a and b. More precisely: Let d be the orthogonal to the midline between a and b in the plane containing the two vectors, let λ^d be the spin component along d and let g^d be the group element changing sign of λ^d. Then $g_{ab} = g^d$.

The case with one orbit and $c = 1/2$ corresponds to electrons and other spin $1/2$ particles. The direction defined by $a = 0$ is some arbitrary fixed direction.

In general, the assumptions of this section may be motivated in a similar manner: First, a conceptual variable $\zeta^a = \zeta^a(\phi)$ is introduced for each a through a chosen focusing, and a suitable group acting on ζ^a is defined. Then λ^a is defined as a reduction of ζ^a to a set of orbits of this group. The essence of Assumption 4.1 is that it is *this* λ^a which is maximally accessible. *This may be regarded as the quantum hypothesis.*

This reasoning works for variables like spin and angular momentum, in general for many discrete e-variables. For theoretical position ξ and theoretical momentum π of a particle, let $\phi = (\xi, \pi)$. Then one can again introduce groups and group elements, and the assumptions of this section are except Assumption 4.2b) are satisfied for this case. A special discussion of continuous e-variables is carried out in Sect. 5.2 below.

4.3 The Toy Model of Spekkens

Nearly since its introduction in the beginning of the last century, discussions of the interpretation of quantum mechanics have taken place. In particular, researchers have disagreed on how the quantum state should be interpreted. Should it be seen as a real state of nature (the ontic view) or does it only represent our knowledge of some focused aspect of nature (the epistemic view)? In my opinion, some synthesis here should be sought, but one should start with an observer and the epistemic process connected to this observer in his particular context. This will give an easy interpretation of the collapse of the wave packet during measurement, and it will also solve paradoxes like that of Schrödinger's cat and that of Wigner's friend. These aspects will be further discussed later in the book after Born's formula and the Schrödinger equation have been introduced and motivated from my point of view. The ontic interpretation arises in my world view from a hypothetical situation where all potential observers communicate and arrive at a common context.

As it stands now, however, the quantum community is divided. Recently there has appeared in the literature certain no-go theorems which seem to support the ontic view. All these theorems are deeply founded, but they rely on certain assumptions. Under these assumptions they show that a pure epistemic view leads to inconsistencies. In particular Pusey et al. (2012) take as a point of departure a certain assumption of separability. This is weakened by Hall (2011) to an assumption of compatibility. Hardy (2012) introduced a different assumption of ontic indifference. The common denominator of these papers is that they show that under the specific assumptions the probability distribution over the ontic states corresponding to different quantum states cannot overlap. See also my discussion of the PBR theorem in Sect. 1.4. A crucial assumption is that the properties of the system can be defined by some state concept.

The toy model of Spekkens (2007) is based on a principle that restricts the amount of knowledge an observer can have about reality. A wide variety of quantum phenomena were found to have analogues within this toy theory, and this can be taken as an argument in favour of the epistemic view of quantum states.

In the simplest version of the toy model, we have one elementary system. This system can be in one of the four ontic states 1, 2, 3 or 4, but our knowledge of this is in principle restricted. We can only know one of the following six epistemic states: (a) The ontic state is 1 or 2; (b) it is 3 or 4; (c) it is 1 or 3; (d) it is 2 or 4; (e) it is 1 or 4; or (f) it is 2 or 3. These are the epistemic states of maximal knowledge.

The ontic base of the state (a) is $\{1, 2\}$ etc.. If the intersection of the ontic bases of a pair of epistemic is empty, then those states are said to be disjoint. Thus (a) and (b) are disjoint, (c) and (d) are disjoint, and (e) and (f) are disjoint. There is a correspondence with certain basis vectors of the two-dimensional complex Hilbert space, where disjointness corresponds to orthogonality in the Hilbert space. For those who knows the Bloch sphere representation of that Hilbert space, the pairs of disjoint epistemic states can be pictured on the intersections of three orthogonal axes with that sphere.

Transformations of the epistemic states correspond to permutations of the ontic state. Thus the underlying group is the permutation group of four symbols, which has 24 elements. Each permutation induces a map between the epistemic states. In the Hilbert space correspondence, the even permutations correspond to unitary transformations, and the odd permutations correspond to anti-unitary transformations.

In my terminology, the system can be described by an inaccessible conceptual variable ϕ which is a vector whose three components are accessible e-variables:

$$\phi = (\lambda^a, \lambda^c, \lambda^e).$$

Here λ^i is the indicator of the event that the epistemic state is i. Each λ^i takes the value 1 or 0. If $\lambda^a = 1$, say, the ontic state is either 1 or 2; if $\lambda^a = 0$, it is either 3 or 4. A complete knowledge of ϕ is equivalent to a knowledge of the ontic state, which is impossible in the Spekkens toy model.

Each λ^i is a maximal accessible e-variable. The event $\lambda^a = 1$ is taken into the event $\lambda^a = 0$ by the even permutation $g^a = (13)(24)$, written in cycle notation. This together with the identity generates the group G^a. Similarly G^c and G^e are generated. The e-variable λ^c is taken into the e-variable λ^a by the even permutation $g_{ca} = (123)(4)$. This permutation can also be written as $g_{af} = g_{be} = g_{db} = g_{fc} = g_{ed}$ if obvious new e-variables are introduced. Similarly, the group elements g_{ac}, g_{ae}, g_{ea}, g_{ce} and g_{ec} are even permutations. The group G is the group of all even permutations. All assumptions of the maximal symmetrical epistemic setting are satisfied except Assumption 4.2b). Thus the Spekkens toy model can not be seen as a special case of the maximal symmetrical epistemic setting, but the simplest case of the toy model is closely related to this.

The next simplest case of the Spekkens toy model consists of two elementary systems. The main requirement from one system carries over: If one has maximal knowledge, then for every system, at every time, the amount of knowledge one possesses about the ontic state of the system at that time must equal the amount of knowledge one lacks. The following discussion is very brief and presupposes a knowledge of Spekkens (2007). There are sixteen ontic states: $1 \cdot 1$, $1 \cdot 2, \ldots, 4 \cdot 4$. It turns out that the valid epistemic states are of two types: The uncorrelated states exemplified by (a) $1 \cdot 1$, $1 \cdot 2$, $2 \cdot 1$ or $2 \cdot 2$, and the correlated states exemplified by (e) $1 \cdot 1$, $2 \cdot 2$, $3 \cdot 3$ or $4 \cdot 4$.

Turning to my terminology, the state (a) can be represented by the event $\lambda^a = 1$, where λ^a is the indicator of the epistemic state (a), an e-variable. The event $\lambda^a = 0$ does not represent an epistemic state of maximal knowledge, however, so the following trick is called for: Let $\lambda^1 = (\lambda^a, \lambda^b, \lambda^c, \lambda^d)$, where λ^a is the indicator of the epistemic state (a), λ^b is the indicator of the epistemic state (b): $3 \cdot 1$, $4 \cdot 1$, $3 \cdot 2$ or $4 \cdot 2$, λ^c is the indicator of the epistemic state $1 \cdot 3$, $1 \cdot 4$, $2 \cdot 3$ or $2 \cdot 4$, and λ^d is the indicator of the epistemic state $3 \cdot 3$, $3 \cdot 4$, $4 \cdot 3$ or $4 \cdot 4$. Allow λ^1 to take the values $(1, 0, 0, 0)$, $(0, 1, 0, 0)$, $(0, 0, 1, 0)$ or $(0, 0, 0, 1)$. Then λ^1 is a maximal e-variable, and the events $\lambda^1 = u_k$ are all valid epistemic states in the Spekkens toy model. A similar trick can be made for the correlated epistemic states.

In Spekkens (2007) transformations between the epistemic states are discussed in terms of permutations between the ontic states. It turns out that there are permutations that take each of $\lambda^a, \lambda^b, \lambda^c, \lambda^d$ and λ^e (with an obvious definition of the last one) into each single other of this set. The transformations taking λ^1 into similar other maximal e-variable are transformations of four-vectors whose components are transformed by permutations. The transformations g^1 etc. can be written in a similar form, though they at the outset only are given by permutations of the components of the four-vector. Thus the group G is a subgroup of the group of transformations of four-vectors whose components are transformed by permutations of 16 elements, that is, a finite group. The assumptions of the maximal symmetrical epistemic setting are satisfied except Assumption 4.2b).

The fact that the Spekkens toy model has a valid epistemic interpretation and is strongly related to many phenomena of quantum mechanics, together with the fact that there is a link between this model and a modification of the maximal symmetrical epistemic setting, gives a strong indication that eventual logical difficulties in the interpretation of the maximal symmetrical epistemic setting and relating it to quantum mechanics, can be overcome. In the next section the formal apparatus of quantum mechanics will be reproduced from the maximal symmetrical epistemic setting under a certain technical condition.

4.4 The Hilbert Space Formulation

Take again as a point of departure the maximal symmetrical epistemic setting. The crucial step now towards the formalism of quantum mechanics is to define a Hilbert space, that is, a complete inner product space which serves as a state space in the formalism (see Appendix B). In ordinary quantum mechanics all observables are identified with operators on such a Hilbert space and every state is identified with a unit vector in the Hilbert space or more generally with a ray proportional to a unit vector. There is a large abstract general theory on this, well known to physicists, but largely unknown to statisticians and many other professionals. My goal here is to rederive this theory from the assumptions of the maximal symmetrical epistemic setting and possibly further assumptions. This may serve as introducing other scientists to the theory. The section is somewhat technical. It can be skimmed at the first reading, but it is in some sense essential for what I feel should be a way to understand ordinary quantum theory. However, it must only be seen as one out of several possible approaches towards the Hilbert space formalism. Several deep theories with a similar purpose exist; see references in Sect. 4.1.

4.4.1 Quantum Reconstruction

Fix $0 \in \mathcal{A}$, and let H be the Hilbert space

$$H = \{f \in L^2(\Phi, \rho) : \ f(\phi) = \tilde{f}(\lambda^0(\phi)) \text{ for some } \tilde{f}\}.$$

Here $L^2(\Phi, \rho)$ is the set of all complex functions f on Φ such that $\int_\Phi |f(\phi)|^2 d\rho < \infty$. Two functions f_1 and f_2 are identified if $\int_\Phi |f_1(\phi) - f_2(\phi)|^2 d\rho = 0$. From now on I will assume that the λ^a's are discrete. Then H is separable. If the λ^a's take d different values, H is d-dimensional. Since all separable Hilbert spaces are isomorphic, it is enough to arrive at the quantum formulation on this H.

Lemma 4.1 *The values u_k^a of λ^a can always be arranged such that $u_k^a = u_k$ is the same for each a ($k = 1, 2, \ldots$).*

Proof By Assumption 4.1

$$\{\phi : \lambda^b = u_k^b\} = \{\phi : \lambda^a(g_{ab}\phi) = u_k^b\} = g_{ba}(\{\phi : \lambda^a(\phi) = u_k^b\}).$$

The sets in brackets on the lefthand side here are disjoint with union Φ. But then the sets in brackets on the righthand side are disjoint with union $g_{ab}(\Phi) = \Phi$, and this implies that $\{u_k^b\}$ gives all possible values of λ^a.

Now I am able to formulate the main result of this section. In the next subsection, I will prove this result under an additional technical assumption. An open question is to find exactly the conditions under which this theorem is valid.

Theorem 4.1

a) *For every a, u_k and associated with every indicator function $I(\lambda^a(\phi) = u_k)$ there is a vector $|a; k\rangle \in H$. The mapping $I(\lambda^a(\phi) = u_k) \to |a; k\rangle$ is invertible in the sense that $|a; k\rangle \neq |b; j\rangle$ for all a, b, j, k except in the trivial case $a = b, j = k$. This inequality is interpreted to mean that there is no phase factor $e^{i\gamma}$ such that $|a; k\rangle = e^{i\gamma}|b; j\rangle$.*
b) *For each a the vectors $|a; k\rangle$ form an orthonormal basis for H.*

This gives us the possibility to interpret the vectors $|a; k\rangle$, corresponding to the indicators $I(\lambda^a(\phi) = u_k)$, as follows:

(1) *The question 'What is the value of λ^a?' has been focused on.*
(2) *Through an epistemic process we have obtained the answer '$\lambda^a = u_k$'.*

When a ket vector is defined, a corresponding bra vector can be defined. The operator corresponding to λ^a can be defined as

$$A^a = \sum_k u_k |a; k\rangle\langle a; k|.$$

In the maximal setting this has non-degenerate eigenvalues.

4.4.2   Proof Under an Extra Assumption

Let U be the left regular representation of G on $L^2(\Phi, \rho)$: $U(g)f(\phi) = f(g^{-1}\phi)$. It is well known that this is a unitary representation. We will seek a corresponding representation of G on the smaller space H.

In the following, recall that upper indices as in g^a indicate variables related to a particular λ^a, here a group element of G^a. Also recall that 0 is a fixed index in $\mathcal{A}$. Lower indices as in g_{ab} has to do with the relation between two different λ^a and λ^b.

Proposition 4.2

a) A (multivalued) representation V of G on the Hilbert space H can always be found.

b) There is an extended group G' such that V is a univalued representation of G' on H.

c) There is a homomorphism $G' \to G^0$ such that $V(g') = U(g^0)$. If $g' \neq e'$ in G', then $g^0 \neq e$ in G^0.

Proof

a) For each a and for $g^a \in G^a$ define $V(g^a) = U(g_{0a})U(g^a)U(g_{a0})$. Then $V(g^a)$ is an operator on H, since it is equal to $U(g_{0a}g^a g_{a0})$, and $g_{0a}g^a g_{a0} \in G^0 = g_{0a}G^a g_{a0}$. For a product $g^a g^b g^c$ with $g^a \in G^a$, $g^b \in G^b$ and $g^c \in G^c$ we define $V(g^a g^b g^c) = V(g^a)V(g^b)V(g^c)$, and similarly for all elements of G that can be written as a finite product of elements from different subgroups.

Let now g and h be any two elements in G such that g can be written as a product of elements from G^a, G^b and G^c, and similarly h (the proof is similar for other cases.) It follows that $V(gh) = V(g)V(h)$ on these elements, since the last factor of g and the first factor of h either must belong to the same subgroup or to different subgroups; in both cases the product can be defined by the definition of the previous paragraph. In this way we see that V is a representation on the set of finite products, and since these generate G by Assumption 4.2b) it is a representation of G.

Since different representations of g as a product may give different solutions, we have to include the possibility that V may be multivalued.

b) Assume as in (a) that we have a multivalued representation V of G. Define a larger group G' as follows: If $g^a g^b g^c = g^d g^e g^f$, say, with $g^k \in G^k$ for all k, we define $g'_1 = g^a g^b g^c$ and $g'_2 = g^d g^e g^f$. Let G' be the collection of all such new elements that can be written as a formal product of elements $g^k \in G^k$. The product is defined in the natural way, and the inverse by for example $(g^a g^b g^c)^{-1} = (g^c)^{-1}(g^b)^{-1}(g^a)^{-1}$. By Assumption 4.2b), the group G' generated by this construction must be at least as large as G. It is clear from the proof of a) that V also is a representation of the larger group G' on H, now a one-valued representation.

c) Consider the case where $g' = g^a g^b g^c$ with $g^k \in G^k$. Then by the proof of a):

$$V(g') = U(g_{0a})U(g^a)U(g_{a0})U(g_{0b})U(g^b)U(g_{b0})U(g_{0c})U(g^c)U(g_{c0})$$

$$= U(g_{0a}g^a g_{a0}g_{0b}g^b g_{b0}g_{0c}g^c g_{c0}) = U(g^0),$$

where $g^0 \in G^0$. The group element g^0 is unique since the decomposition $g' = g^a g^b g^c$ is unique for $g' \in G'$. The proof is similar for other decompositions. By the construction, the mapping $g' \to g^0$ is a homomorphism.

Assume now that $g^0 = e$ and $g' \neq e'$. Since $U(g^0)\tilde{f}(\lambda^0(\phi)) = \tilde{f}(\lambda^0((g^0)^{-1}(\phi)))$, it follows from $g^0 = e$ that $U(g^0) = I$ on H. But then from what has been just proved, $V(g') = I$, and since V is a univariate representation, it follows that $g' = e'$, contrary to the assumption.

Assumption 4.3

a) U is an irreducible representation of every cyclic subgroup of the group $\tilde{G}^0$ on H other than the trivial group, and the dimension d of H is larger or equal to 2.

b) The representation V of the whole group G is really multivalued on the elements g_{ab}.

c) When finding this basis, one can choose $\tilde{f}_k$ and $\tilde{f}_j$ in such a way that there exists a λ_1 such that $\tilde{f}_k(\tilde{g}^0\lambda_1) \neq \tilde{f}_j(\lambda_1)$ for all $\tilde{g}^0 \epsilon \tilde{G}^0$ in the sense that the two sides can not be made equal by introducing a phase factor.

Now choose an orthonormal basis for H: $f_1, \ldots, f_d$ where $f_k(\phi) = \tilde{f}_k(\lambda^0(\phi))$, and where the interpretation of f_k is that $\lambda^0 = u_k$. Write $|0; k\rangle = f_k(\phi)$.

Lemma 4.2 *For every k and every $g^0 \in G^0$, $g^0 \neq e$, we have $U(g^0)f_k \neq f_k$ in the sense that the two functions can not be made equal by multiplying with a phase factor.*

Proof Let $d \geq 2$. Assume that there exist γ, k and $g^0 \neq e$ such that $U(g^0)f_k = e^{i\gamma}f_k$. Then $e^{i\gamma/2}f_k$ span a one-dimensional subspace of H which is invariant under the cyclic group generated by g^0, contrary to the assumption of irreducibility.

Introduce now the assumption that the representation V really is multivalued. Let g'_{0a1} and g'_{0a2} be two different elements of the group G', both corresponding to g_{0a} of G. Define $g'_a = (g'_{0a1})^{-1}g'_{0a2}$. Then $g'_a \neq e'$ in G'. By the homomorphism of Proposition 4.2c), let $g'_a \to g^0_a$. Then $g^0_a \neq e$ in G^0. Now define

$$|a; k\rangle = \tilde{f}_k(\lambda^0(g^0_a\phi)) = U((g^0_a)^{-1})|0; k\rangle.$$

Proof of Theorem 4.1 Under Assumption 4.3 By Lemma 4.2, $|a; k\rangle \neq |0; k\rangle$. Here and below, inequality of state vectors is interpreted to mean that they can not be made equal by introducing a phase factor.

Next let $j \neq k$. I will prove that the basis functions $f_1, \ldots, f_d$ can be chosen so that $|a; k\rangle \neq |0; j\rangle$ for all a. To this end, choose $\tilde{f}_j$ and $\tilde{f}_k$ in such a way that there exists an λ^j_0 such that $\tilde{f}_k(\tilde{g}^0\lambda^j_0) \neq \tilde{f}_j(\lambda^j_0)$ for all $\tilde{g}^0 \epsilon \tilde{G}^0$. Then for any fixed g, $\tilde{f}_{kg}$ defined by $\tilde{f}_{kg}(\lambda^0(\phi)) = \tilde{f}_k(\lambda^0(g\phi))$ is different from $\tilde{f}_j$, and $|a; k\rangle \neq |0; j\rangle$ for all a.

The proof that $|a; k\rangle \neq |b; j\rangle$ (except in the trivial case $a = b, k = j$) holds under Assumption 4.3, is a straightforward extension.

The vectors $|0; k\rangle$ are chosen to be an orthonormal basis for H. Since $|a; k\rangle = U|0; k\rangle$ for some unitary U, it follows that the vectors $|a; k\rangle$ form an orthonormal basis.

Several authors have approached the foundation of quantum mechanics through group representation theory, one example being Mirman (1995), another Smilga (2017).

Assumption 4.3 is not satisfied for the spin/ angular momentum case. For this case the group $\tilde{G}^0$ is too small to generate all states $|a; k\rangle$ from $|0; k\rangle$. I will show directly in the next chapter that Theorem 4.1 holds in general for the case of spin/ angular momentum (Corollary 5.1 of Sect. 5.2). But this also means that it must be possible to weaken Assumption 4.3) in some way.

Theorem 4.1, saying that the question-and-answer pairs can be put in one-to-one correspondence with some states in a concrete Hilbert space, must hold under a set of assumptions including the spin/ angular momentum case.

In the qubit case (Hilbert space dimension 2; answers to questions: -1 and $+1$), the question-and-answer pairs can be put in one-to-one correspondence with *all* state vectors in the Hilbert space. (Sect. 5.1.1 and Proposition 5.4 of Sect. 5.2.)

4.4.3   The Interpretation Argument

In the previous subsections I started with a setting where questions-and-answers could be defined, and arrived at Hilbert space unit vectors. Now start by assuming the Hilbert space formulation, and let $|\psi\rangle$ be an arbitrary unit vector in the Hilbert space H. This vector is of course the eigenvector of many operators in H. Assume that one can find such an operator A such that

1. A is physically meaningful, i.e., can be associated by an e-variable λ.
2. $|\psi\rangle$ is an eigenvector corresponding to a non-degenerate eigenvalue u of A.

Then $|\psi\rangle$ can be interpreted as a question 'What is the value of λ?' together with a definite answer '$\lambda = u$'. The non-degenerate eigenvalue case corresponds to the *maximal* epistemic setting.

4.5   The General Symmetrical Epistemic Setting

Go back to the definition of the maximal symmetrical epistemic setting. Let again ϕ be the inaccessible conceptual variable and let $\lambda^a = \lambda^a(\phi)$ for $a \in \mathcal{A}$ be the maximal accessible conceptual variables satisfying Assumption 4.1. Let the corresponding induced groups G^a and G satisfy Assumptions 4.2 and 4.3 (or some weaker assumption which can replace Assumption 4.3). Finally, let t^a for each a be an arbitrary function on the range of λ^a, and assume that we instead of focusing on λ^a, focus on $\theta^a = t^a(\lambda^a)$ for each $a \in \mathcal{A}$. I will call this the symmetrical epistemic setting; the e-variables θ^a are no longer maximal.

Let the Hilbert space be as in Sect. 4.4.1 with an orthonormal basis redefined to be $\{|a'; j\rangle\}$ for each a. Let $\{u_j\}$ be the values of λ^a, and let u_k^a be the values of

$\theta^a = t^a(\lambda^a)$. Define $C_k^a = \{j : t^a(u_j) = u_k^a\}$, let V_k^a be the space spanned by $\{|a'; j\rangle; j \in C_k^a\}$ and let Π_k^a be the projection upon V_k^a. Finally, let $|a; k\rangle$ be any unit vector in V_k^a.

Interpretation of the State Vector $|a; k\rangle$ *(1) The question: 'What is the value of θ^a?' has been posed. (2) We have obtained the answer $\theta^a = u_k^a$. Both the question and the answer are contained in the state vector.*

From this we may define the operator connected to the e-variable θ^a:

$$A^a = \sum_k u_k^a \Pi_k^a = \sum_j t^a(u_j)|a'; j\rangle\langle a'; j|. \tag{4.1}$$

Then A^a is no longer necessarily an operator with distinct eigenvalues, but A^a is still Hermitian: $A^{a\dagger} = A^a$.

Interpretation of the Operator A^a *This gives all possible states and all possible values corresponding to the accessible e-variable θ^a.*

The general decomposition $A^a = \sum_k u_k^a \Pi_k^a$ will be important in Sect. 5.4 and Sect. 5.7, and will be further discussed there.

The projectors $|a; k\rangle\langle a; k|$ and hence the ket vectors $|a; k\rangle$ are no longer uniquely determined by A^a: They can be transformed arbitrarily by unitary transformations in each space corresponding to one eigenvalue. As long as the focus is only on θ^a, or A^a, I will redefine $|a; k\rangle$ by allowing it to be subject to such transformations. These transformed eigenvectors all still correspond to the same eigenvalue, that is, the same observed value of θ^a and they give the same operators A^a. In particular, in the maximal symmetric epistemic setting I will allow an arbitrary constant phase factor in the definition of the $|a; k\rangle$'s.

A more precise state interpretation is then to let the whole vector space of such transformed vectors $|a; k\rangle$ represent a question-and-answer pair. This will be gone more thoroughly into in Sect. 5.7.3.

As an example of the general construction, assume that λ^a is a vector: $\lambda^a = (\theta^{a1}, \ldots, \theta^{am})$. The different θ's may be connected to different subsystems. This example is highly relevant when considering several observers. One single observer may have access to just a few subsystems. In addition he has his own context. From this context he may define his accessible and inaccessible conceptual variables. In the same way a group of observers may through verbal communication arrive at a common context, and from this context one may define their accessible and inaccessible conceptual variables. Assume that these observers together observe a particular physical system.

Assumption 4.4 *For a given physical system at some particular time one can imagine an observer or a group of communicating observers for which the assumptions of the symmetrical epistemic setting are satisfied. In some cases all possible observers agree on the physical observations, and these then describe an objective property of the system.*

So far I have kept the same groups G^a and G when going from λ^a to $\theta^a = t^a(\lambda^a)$, that is from the maximal symmetrical epistemic setting to the general symmetrical epistemic setting. This implies that the (large) Hilbert space will be the same. A special case occurs if t^a is a reduction to an orbit of G^a. This is the kind of model reduction discussed in Sect. 2.2. Then the construction of the previous sections can also be carried with a smaller group action acting just upon an orbit, resulting then in a smaller Hilbert space. In the example $\lambda^a = (\theta^{a_1}, \ldots, \theta^{a_m})$. it may be relevant to consider one Hilbert space for each subsystem. Then one can write a state vectors corresponding to λ^a as

$$|a; k\rangle = |a_1; k_1\rangle \otimes \ldots \otimes |a_m; k_m\rangle$$

in an obvious notation, where $a = (a_1, \ldots, a_m)$ and $k = (k_1, \ldots, k_m)$. The large Hilbert space is however the correct space to use when the whole system is considered. In this Hilbert space the subsystem ket vectors will have degenerate eigenvalues and correspond to the general symmetrical epistemic setting.

At any time we can also imagine non-communicating observers. Then for each particular observers the assumptions of the general symmetrical setting may be assumed to apply. Particular state vectors in each observer's Hilbert space may be linear combinations of primitive state vectors in the form of a tensor product. These are called entangled states when they can not be reduced to a primitive form, and play an important role in many areas of quantum physics.

Assumption 4.4 is assumed for a large class of physical systems. Through the imagined observers the constructions of this chapter may be carried out, and for each case a Hilbert space may e constructed.

This is the connection between my theory and the formal quantum theory defined in textbooks. I will claim that the theory defined by having the maximal symmetrical epistemic setting as a point of departure is from one point of view more intuitive than the ordinary formal theory.

References

Ballentine, L. E. (1998). *Quantum mechanics. A modern development.* Singapore: World Scientific.

Bell, J. S. (1987). *Speakable and unspeakable in quantum mechanics.* Cambridge: Cambridge University Press.

Casinelli, G., & Lahti, P. (2016). An axiomatic basis for quantum mechanics. *Foundations of Physics, 46,* 1341–1373.

Caves, C. M., Fuchs, C. A., Schack, R. (2002). Quantum probabilities as Bayesian probabilities. *Physical Review, A65,* 022305.

Chiribella, G., D'Ariano, G. M., Perinotti, P. (2010). Informational derivation of quantum theory. arXiv: 1011.6451 [quant-ph].

Einstein, A., Podolsky, B., & Rosen, N. (1935). Can quantum-mechanical description of physical reality be considered complete? *Physical Review, 47,* 777–780.

Fields, C. (2011). Quantum mechanics from five physical assumptions. arXiv: 1102.0740 [quant-ph].

Fivel, D. I. (2012). Derivation of the rules of quantum mechanics from information-theoretic axioms. *Foundations of Physics, 42*, 291–318.

Fuchs, C. A. (2002). Quantum mechanics as quantum information (and only a little more). In A. Khrennikov (Ed.), *Quantum theory: Reconsideration of foundations.* Växjö: Växjö University Press.

Fuchs, C. A. (2010). QBism, the Perimeter of Quantum Bayesianism. arXiv: 1003.5209v1 [quant-ph].

Fuchs, C. A., & Peres, A. (2000). Quantum theory needs no interpretation. *Physics Today, S-0031-9228-0003-230-0*; Discussion *Physics Today, S-0031-9228-0009-220-6.*

Fuchs, C. A., Mermin, N. D., & Schack, R. (2013). An introduction to QBism with an application to the locality of quantum mechanics. arXiv: 1311.5253v1 [quant-ph].

Fuchs, C. A., & Schack, R. (2011). A quantum-Bayesian route to quantum-state space. *Foundations of Physics, 41*, 345–356.

Hall, M. J. W. (2011). Generalizations of the recent Pusey-Barrett-Rudolph theorem for statistical models of quantum phenomena. arXiv: 1111.6304 [quant-ph].

Hardy, L. (2001). Quantum theory from five reasonable axioms. arXiv: 0101012v4.[quant-ph].

Hardy, L. (2012). Are quantum states real? arXiv: 1205.1439 [quant-ph].

Helland, I. S. (2006). Extended statistical modeling under symmetry; the link toward quantum mechanics. *Annals of Statistics, 34*, 42–77.

Helland, I. S. (2008). Quantum mechanics from focusing and symmetry. *Foundations of Physics, 38*, 818–842.

Helland, I. S. (2010). *Steps towards a unified basis for scientific models and methods.* Singapore: World Scientific.

Kuhlmann, M. (2013). What is real? *Scientific American, 309*(2), 32–39.

Masanes, L. (2010). Quantum theory from four requirements. arXiv: 1004.1483 [quant-ph].

Mirman, R. (1995). *Group theoretical foundations of quantum mechanics.* Lincoln, NE: iUniverse.

Nisticò, G., & Sestito, A. (2011). Quantum mechanics, can it be consistent with locality? *Foundations of Physics, 41*, 1263–1278.

Pusey, M. F., Barrett, J., & Rudolph, T. (2012). On the reality of quantum states. *Nature Physics, 8*, 475–478.

Schack, R. (2006). Bayesian probability in quantum mechanics. In *Proceedings of Valencia/ ISBA World Meeting on Bayesian Statistics.*

Searle, S. R. (1971). *Linear models.* New York: Wiley.

Smilga, W. (2017). Towards a constructive foundation of quantum mechanics. *Foundations of Physics, 47*, 149–159.

Spekkens, R. W. (2007). In defense of the epistemic view of quantum states: A toy theory. *Physical Review A, 75*, 032110.

Timpson, C. G. (2008). Quantum Bayesianism: A study. *Studies in history an philosophy of modern physics, 39*, 579–609.

von Neumann, J. (1932). *Mathematische Grundlagen der Quantenmechanik.* Berlin: Springer.

Chapter 5
Aspects of Quantum Theory

Abstract The basic formalism of quantum theory is reviewed and put into my setting. In particular, spin and angular momentum are considered, and also e-variables like position and momentum. For the spin/angular momentum case, the correspondence between question-and-answer pairs and unit Hilbert space vectors is proved directly. A link to statistical inference is found by proving a focused version of the likelihood principle. From this and from an assumption of rationality the Born rule is proved. A macroscopic example is proposed. Measurements and quantum statistical inference are briefly discussed. Bell's inequality issues are considered from an epistemic point of view. The free will theorem is mentioned, and the Schrödinger equation is derived in the one-dimensional case. Several so-called paradoxes are explained from an epistemic setting. Quantum measurements when the density matrix is completely unknown are discussed from a statistical point of view. A discussion of the book so far concludes the chapter.

5.1 Basic Quantum Theory

In this section I will no longer assume the symmetrical epistemic setting, and we thus will for a moment forget the group-theoretical assumptions like Assumptions 4.1–4.3. I just take as a point of departure a finite dimensional complex vector space H with kets $|k\rangle$ being orthonormal basis vectors, and corresponding bras $\langle k|$. The one-dimensional predictors $|k\rangle\langle k|$ are defined as before, and all operators on H are of the form

$$A = \sum_k u_k |k\rangle\langle k| \text{ with } \langle i|j\rangle = \delta_{ij}. \tag{5.1}$$

This is equivalent to defining the action of A on any vector of H:

$$A \sum_i a_i |i\rangle = \sum_i a_i u_i |i\rangle. \tag{5.2}$$

© Springer-Verlag GmbH Germany, part of Springer Nature 2018

I. S. Helland, *Epistemic Processes*, https://doi.org/10.1007/978-3-319-95068-6_5

A fundamental assumption of quantum theory is that there to each physical system correspond a Hilbert space H, and that there to each simple e-variable/ observable θ of this physical system correspond an operator A of this Hilbert space. Basic operator theory is reviewed in Appendix B.

Assume first that A has distinct eigenvalues. This corresponds to the maximal symmetric setting of Sect. 4.2. Then all eigenvectors of A define quantum states. In the general case, we can extend A by taking into account auxiliary eigenvalues in such a way that the extended operator is maximal, and let the states be the eigenvectors of this extended operator; see below. All these states can be given an interpretation in terms of a focused question: 'What is the value of θ?' together with a definite answer: '$\theta = u_k$'. Often the discussion will involve several e-variables $\theta^a, \theta^b, \ldots$. The state vector corresponding to $\theta^a = u_k^a$ will then be called $|a; k\rangle$.

Given a ket vector $|k\rangle$ there is an infinity of operators A which have $|k\rangle$ as an eigenvector. In many cases there is at least one such operator which is physically meaningful, i. e., is associated with some e-variable θ. In some cases, e.g., the spin of an electron, this operator is unique in a specific sense.

From this all the features of elementary quantum mechanics follow except the probability statements and the time development of states, which I will come back to later. Quantum states that are not directly constructed as eigenvectors of any physically meaningful operator can be found via the time dependent Schrödinger equation; see Sect. 5.11, and also in some cases by considering physically meaningful linear combination of state vectors; see below.

The operators can be multiplied as also discussed in Appendix B. The multiplication is associative, but not commutative. As usual we define the commutator as

$$[A, B] = AB - BA.$$

The Hermitian adjoint operator $A^\dagger$ is defined such that the ket $A^\dagger|k\rangle$ corresponds to the bra $\langle k|A$, in other words $\langle i|A^\dagger j\rangle = \langle iA|j\rangle$ for all $|i\rangle$, $|j\rangle$. This $A^\dagger$ can also be defined by complex conjugating the eigenvalues in the formula above. The observables are defined as the Hermitian operators: $A^\dagger = A$, that is, the case where the eigenvalues u_k are real. In general one has $(AB)^\dagger = B^\dagger A^\dagger$. The possible values of the observables are their eigenvalues, and the states are given by the ket vectors.

As already stated, the maximal symmetrical epistemic setting of the previous chapter correspond to the case where the eigenvalues of the operator A are distinct, but this is generalized in the general symmetrical setting of Sect. 4.5. There I also had a brief discussion on the choice of Hilbert space. A smaller Hilbert space is relevant when only a subsystem is considered, but when the whole system is to be described, a larger Hilbert space is needed.

These considerations are highly relevant when considering several observers. One single observer may have access to just a few subsystems. In addition he has his own context. From this context one can define what is his accessible and inaccessible conceptual variables. In the same way a group of several observers may through verbal communication arrive at a common context, and from this context one can define their accessible and inaccessible conceptual variables. Imagine that these observers together observe a particular physical system, and consider the corresponding Hilbert space.

Assumption 5.4 *For a given physical system at some particular time one can either imagine an observer or a group of communicating observers for which assumptions leading to the Hilbert space formulation are satisfied. In many cases all real and imagined observers agree on the physical observations, in which case this is considered part of the ontic world.*

It is important that for every physical system we can consider imagined observers. All real and imagined observers may agree on observations like charge or non-relativistic mass of a particle. For these observations we do not need the formalism of quantum mechanics. For other variables the construction of this section applies, leading to a formal apparatus which is identical to that of quantum theory.

At any time we can also imagine non-communicating observers. For each observer then the general setting corresponding to the quantum formulation may apply. Particular state vectors in each observer's Hilbert space may then be linear combinations of primitive state vectors of the form $|a_1; k_1\rangle \otimes \ldots \otimes |a_s; k_s\rangle$. When these can not be reduced to a primitive form, they are called entangled state vectors. Entangled state vectors play an important role in many discussions of quantum mechanics, and I will come back to them briefly below.

When the physical system available to an observer A is subject to a non-maximal setting resulting in an operator $A^a = \sum_k u_k^a |a; k\rangle\langle a; k|$, we can always imagine another observer B, communicating with A, with an operator $B^b = \sum_j v_j^b |b; j\rangle\langle b; j|$, so that the operator $A^a \otimes B^a = \sum_{k,j} u_k^a v_j^b (|a; k\rangle \otimes |b; j\rangle)(\langle a; k| \otimes \langle b; j|)$ corresponds to a maximal setting, i.e., has distinct eigenvalues.

Assumption 5.4 is assumed to hold for any physical system, and also for combinations of physical systems. Through the imagined observers the construction of this section can be carried out, and for each case a Hilbert space can be constructed.

Connected to any general physical system, one may have several e-variables θ and corresponding operators A. There is well-known theorem saying that, in my formulation, $\theta^1, \ldots, \theta^n$ are compatible, that is, there exists an e-variable λ such that $\theta^i = t^i(\lambda)$ for some functions t^i if and only if the corresponding operators commute:

$$[A^i, A^j] \equiv A^i A^j - A^j A^i = 0 \text{ for all } i, j.$$

(See Holevo 2001 or Ballentine 1998.) Compatible e-variables may in principle be estimated simultaneously with arbitrary accuracy.

In an attempt to make all this a little more concrete, look again at Example 4.15. Let J_x, J_y and J_z be the operators corresponding to spin in three orthogonal directions x, y and z. In quantum mechanical texts (see Messiah 1969) it is shown that there is a constant c such that these operators satisfy the commutation relations:

$$[J_x, J_y] = icJ_z, \quad [J_y, J_z] = icJ_x, \quad [J_x, J_z] = icJ_y.$$

This can also be proved fairly easily directly in my setting for the electron spin case $j = 1/2$, using the geometry of SU(2). In standard quantum mechanics $c = \hbar/2$, where $\hbar$ is Planck's constant. We will choose units here such that $c = 1/2$. In great generality, commutation relations may be derived from group properties by exploiting the relation between Lie groups and Lie algebras; see for instance Barut and Raczka (1985).

Several consequences of the above commutation relations are derived in standard texts, for instance Messiah (1969). First it is shown that J_z commutes with $J^2 = J_x^2 + J_y^2 + J_z^2$ and that J^2 has eigenvalues of the form $j(j + 1)$, where j is integer or half integer. It is well known, see the discussion above, that commuting operators can be simultaneously diagonalized. In terms of the corresponding e-variables $(\theta_x, \theta_y, \theta_z)$ this means that the vector $(\|\theta\|^2, \theta_z)$ is accessible, where $\|\theta\|^2 = \theta_x^2 + \theta_y^2 + \theta_z^2$. Given j, the eigenvalues of J_z are of the form $-j, -j+1, \ldots, j-1, j$ as anticipated in Example 4.15. Also, eigenvectors can be explicitly discussed.

We conclude from this that we have two possible situations:

1. J^2 is known; more explicitly, the squared modulus $\|\theta\|^2$ is known, and takes one of the values $j(j + 1)$. Then the situation is exactly as in Example 4.15, in particular the first assumptions of the maximal symmetrical epistemic setting are satisfied.
2. The squared modulus $\|\theta\|^2$ is unknown. Then the operator $J^2 \otimes J_z$ (taking infinitely many, but discrete values) can be diagonalized, can be understood in terms of conceptual variables, but is not directly given in terms of a maximal symmetrical epistemic setting.

In conclusion, the first assumptions from Sect. 4.2 defining a symmetrical epistemic setting are sometimes satisfied, sometimes not for a given quantum mechanical situation, but the introduction of conceptual variables does seem to be useful for understanding what is going on. Model reduction seems to be crucial here.

Let now by a slight change of notation $J = (J_x, J_y, J_z)$ be the inaccessible total angular momentum of a system of particles where $\|J\|^2 = J(J + 1)$ is known. Assume that J is the sum of two spins j_1 and j_2 where $\|j_1\|^2 = j_1(j_1 + 1)$ and $\|j_2\|^2 = j_2(j_2 + 1)$ are known. Let $|m_i\rangle$ be the state where $j_{iz} = m_i$ for $-j_i \leq m_i \leq j_i$. Then the state $|M\rangle$ where $J_z = M$ can be decomposed into

$$|M\rangle = \sum_{m_1 m_2} c_{Mm_1m_2} |m_1\rangle \otimes |m_2\rangle.$$

The coefficients $c_{Mm_1m_2}$, nonzero only for $m_1 + m_2 = M$, are called Clebsch-Gordon coefficients and are discussed in standard quantum mechanical texts like Messiah (1969). Generalizations, only more technically involved, exist when J is the sum of more than two spins or angular momenta. This is an important example of a situation where new states are found by taking linear combinations of a basic set of state vectors.

From elementary quantum mechanical texts one can get the impression that all linear combinations of state vectors in a Hilbert space are possible state vectors. This is however only true under certain qualifications (I will define superselection rules later). Nevertheless, taking linear combinations of state vectors leads to the introduction of interesting and important quantum mechanical phenomena, in particular that of entanglement, which will be touched upon in Sect. 5.8.

Superposition of quantum states can be introduced in my setting as follows: Assume that we know a state $|a; k\rangle$, that is, we know that $\theta^a = u_k^a$ for an e-variable θ^a, and that we are interested in information about another e-variable θ^b. Since $\sum_j |b; j\rangle\langle b; j| = 1$, we have

$$|a; k\rangle = \sum_j |b; j\rangle\langle b; j|a; k\rangle = \sum_j \langle b; j|a; k\rangle|b; j\rangle. \tag{5.3}$$

In general, when the corresponding operators A^a and A^b do not commute, this is a genuine linear combination of states $|b; j\rangle$, and part of the interpretation in my language is that we do not know the value of θ^b. A situation like this can be illustrated by the two-slit experiment, which I will come back to later, and in many other settings.

However; there are also other situations than the one sketched above where linear combinations of state vectors are of interest. In more generality, given a set of basis vectors $\{|j\rangle\}$ for the Hilbert space, new state vectors of interest can be of the form

$$|\psi\rangle = \sum_j c_j |j\rangle, \tag{5.4}$$

where $\{c_j\}$ is an arbitrary set of complex numbers such that $\sum_j |c_j|^2 = 1$.

To assume a state $|a; k\rangle$ is to assume perfect knowledge of the e-variable θ^a: $\theta^a = u_k^a$. Such perfect knowledge is rarely available. In practice we have data z^a about the system, and use these data to obtain knowledge about θ^a. Let us start with Bayesian inference. This assumes prior probabilities π_k^a on the values u_k^a, and after the inference we have posterior probabilities $\pi_k^a(z^a)$. In either case we summarize this information in the density operator:

$$\rho^a = \sum_k \pi_k^a |a; k\rangle\langle a; k|.$$

Interpretation of the Density Operator ρ^a *(1) We have posed the question 'What is the value of θ^a?' (2) We have specified a prior or posterior probability distribution π_k^a over the possible answers. The probability for all possible answers to the question, formulated in terms of state vectors, can be recovered from the density operator.*

A third possibility for the probability specifications is a confidence distribution; see Sects. 2.1.3 and 2.1.4 and references there. For discrete θ^a the confidence distribution function H^a is connected to a discrete distribution, which gives the probabilities π_k^a. Extending the argument of Xie and Singh (2013) to this situation, this should not be looked upon as a distribution *of* θ^a, but a distribution *for* θ^a, to be used in the epistemic process.

Since the sum of the probabilities is 1, the trace (sum of eigenvalues, or equivalently, sum of diagonal terms) of any density operator is 1. In the quantum mechanical literature, a density operator is any positive operator with trace 1. Given a set of basis vectors $\{|k\rangle\}$, a density operator is any operator of the form

$$\rho = \sum_k \pi_k |k\rangle \langle k|,$$

where $\sum_k \pi_k = 1$.

Lemma 5.3 *A density operator corresponds to a pure state if and only if the trace of its square is 1.*

Proof $\mathrm{trace}((\rho)^2) = \sum_k (\pi_k)^2 \leq \sum_k \pi_k = 1$. Equality can hold here if and only if only one π_k is 1 and the rest 0. Then ρ is a one-dimensional projector, corresponding to a single ket-vector.

A good modern introduction to quantum mechanics, in particular to the Hilbert space formalism, is Ballentine (1998). Concepts and methods of quantum mechanics are discussed in Peres (1993), and also in other, more recent books.

5.1.1 The Qubit Case; The Case of Dimension 2

Let us look closer on the case where the Hilbert space dimension is 2. One particular case of this is the spin 1/2 particle, say the spin components of an electron. The discussion here often starts with the Pauli spin operators

$$\sigma_x = \begin{pmatrix} 0 & 1 \\ 1 & 0 \end{pmatrix}, \quad \sigma_y = \begin{pmatrix} 0 & -i \\ i & 0 \end{pmatrix}, \quad \sigma_z = \begin{pmatrix} 1 & 0 \\ 0 & -1 \end{pmatrix}. \tag{5.5}$$

All of these have eigenvalues $+1$ and -1, the possible spin components in the x-direction, y-direction and z-direction, respectively. The corresponding eigenvectors are $(1\ 1)'$ and $(1\ -1)'$ for σ_x, $(i\ -1)'$ and $(i\ 1)'$ for σ_y and $(1\ 0)'$ and $(0\ 1)'$ for σ_z. In an obvious way, each of these corresponds to a question-and-answer pair for spin component.

The Pauli spin operators are Hermitian and satisfy

$$\text{trace}(\sigma_x) = \text{trace}(\sigma_y) = \text{trace}(\sigma_z) = 0;$$

$$\sigma_x^2 = \sigma_y^2 = \sigma_z^2 = I;$$

$$\sigma_x\sigma_y = -\sigma_y\sigma_x = i\sigma_z, \ \sigma_y\sigma_z = -\sigma_z\sigma_y = i\sigma_x, \ \sigma_z\sigma_x = -\sigma_x\sigma_z = i\sigma_y.$$

Also, the four matrices I, σ_x, σ_y and σ_z span all Hermitian matrices, and one can define the density operator

$$\rho^r = \frac{1}{2}(I + r_x\sigma_x + r_y\sigma_y + r_z\sigma_z) = \frac{1}{2}(I + \mathbf{r} \cdot \boldsymbol{\sigma}). \tag{5.6}$$

This is the most general density operator, i.e., operator with $\text{trace}(\rho^r) = 1$. A straightforward calculation gives $\text{trace}((\rho^r)^2) = \frac{1}{2}(1 + \|\mathbf{r}\|^2)$. Thus in order that (5.6) shall be a density operator (see the proof of Lemma 5.3), we must have $\|\mathbf{r}\| \leq 1$. The pure states (Lemma 5.3) correspond to $\|\mathbf{r}\| = 1$. This gives a one-to-one correspondence between ket vectors in the two dimensional Hilbert space and real-valued three-dimensional vectors $\|\mathbf{r}\|$ of norm 1. This three-dimensional picture is called the Bloch sphere.

The Bloch sphere representation can also be used to describe measurements. (For more details and for a far-reaching recent elaboration of this; see Aerts et al. 2016.) To this end, focus on the eigenstates of the e-variable that is measured. More precisely, for an e-variable represented by the Hermitian operator $A = u_1 P_1 + u_2 P_2$, where $P_k = |k\rangle\langle k|$ is the projection associated with the ket vector $|k\rangle$, we can write: $P_k = \frac{1}{2}(I + \mathbf{n}_k \cdot \boldsymbol{\sigma})$, where $\mathbf{n}_k$ is the unit vector representing the ket vector $|k\rangle$ on the Bloch sphere, $k = 1, 2$.

The two eigenvectors being orthogonal, we have $P_1 P_2 = 0$, so that $\text{trace}(P_1 P_2) = \frac{1}{2}(1 + \mathbf{n}_1 \cdot \mathbf{n}_2) = 0$, implying $\mathbf{n}_1 = -\mathbf{n}_2$, i.e., the two vectors representative of P_1 and P_2 must point in opposite directions.

In this picture, a ket vector corresponding to a vector $\mathbf{n}$ on the Bloch sphere can be given the interpretation: We have asked the question: 'What is the spin component corresponding to $\mathbf{n}$?' with the answer $+1$. The ket vector corresponding to a vector $-\mathbf{n}$ is given the interpretation: 'What is the spin component corresponding to $-\mathbf{n}$?' with the answer $+1$, or equivalently: 'What is the spin component corresponding to $\mathbf{n}$?' with the answer -1.

In this way we have a one-to-one correspondence between the ket vectors in a two-dimensional Hilbert space and question-and-answer pairs for spin components. This correspondence will be proved in another way in the next section.

5.2 More on Angular Momentum

Since all separable Hilbert spaces of a given dimension are isomorphic, we are free to and will in this section use the ordinary Hilbert space formulation used in textbooks for angular momenta/spins. The discussion here will rely on Messiah (1969). Let the angular momentum operator be $\mathbf{J}$, let j in $\|\mathbf{J}\|^2 = j(j + 1)$ be fixed, let J_x, J_y and J_z be the angular momentum operators in directions x, y and z, respectively, and let $J_a = a_1 J_x + a_2 J_y + a_3 J_z$ be the operator corresponding to angular momentum θ^a in the direction $a = (a_1, a_2, a_3)$. Without loss of generality, assume $\sum_i a_i^2 = 1$.

Proposition 5.3 *For each a and each k ($k = -j, -j + 1, \ldots, j - 1, j$) there is exactly one normalized ket vector $|v\rangle = |a; k\rangle$ with arbitrary phase such that $J_a|v\rangle = k|v\rangle$. This ket vector corresponds to the question 'What is the angular momentum component θ^a in the direction a?' together with the definite answer '$\theta^a = k$'.*

Proof Let $\{|m\rangle\}$; $m = -j, \ldots, j$ be the normalized eigenstates of J_z, and seek a ket vector $|v\rangle = \sum_{m=-j}^{j} b_m |m\rangle$ with $\sum |b_m|^2 = 1$ satisfying $J_a|v\rangle = k|v\rangle$. From Messiah (1969) the operators $J^+ = J_x + iJ_y$ and $J^- = J_x - iJ_y$ satisfy

$$J^+|m\rangle = \sqrt{(j - m)(j + m - 1)}|m + 1\rangle = A_m|m + 1\rangle; \ m \neq j,$$

$$J^-|m\rangle = \sqrt{(j + m)(j - m + 1)}|m - 1\rangle = B_m|m - 1\rangle; \ m \neq -j.$$

Solving this for J_x and J_y leads to

$$J_a|v\rangle = (a_1 J_x + a_2 J_y + a_3 J_z) \sum_{m=-j}^{j} b_m |m\rangle$$

$$= \sum_{m=-j}^{j} [\frac{1}{2}(a_1 - ia_2)b_{m-1}A_{m-1} + \frac{1}{2}(a_1 + ia_2)b_{m+1}B_{m+1} + a_3 b_m m]|m\rangle$$

if we define $A_{-j-1} = B_{j+1} = 0$. Putting this equal to $k \sum b_m |m\rangle$ we get the recursion relations

$$kb_m - a_3 mb_m - \frac{1}{2}(a_1 - ia_2)b_{m-1}A_{m-1} = \frac{1}{2}(a_1 + ia_2)b_{m+1}B_{m+1}$$

for $k, m = -j, \ldots, j$.

Put $b_{-j} = c$. Then the recursion relation first gives $b_{-j+1} = \frac{(k+ja_3)c\sqrt{2}}{(a_1+ia_2)\sqrt{j}}$, and the same relation then determines b_{m+1} for $m = -j + 1, \ldots j - 1$. The relation for $m = j$ gives an eigenvalue equation for k, but we already know from $J^a|v\rangle = k|v\rangle$

that this has solutions $k = -j, -j+1, \ldots, j-1, j$. Finally, $|c|$ is determined from $\sum |b_m|^2 = 1$. Thus, modulo an arbitrary phase factor, we have a unique ket vector $|v\rangle$ determined from the basis vectors $|m\rangle$, $m = -j, \ldots, j$.

Corollary 5.1 *Theorem 4.1 of Sect. 4.4.1 is valid for the spin/angular momentum case.*

Proof This follows here from the construction above, giving a solution depending in a unique way upon $a = (a_1, a_2, a_3)$ on the unit sphere.

Corollary 5.2 *The epistemic ket vectors of Proposition 5.3 form a set in the Hilbert space determined by k and by two independent real parameters.*

Proof These simple epistemic states may be indexed by k and by a_1, a_2 and $a_3 = \pm\sqrt{1 - a_1^2 - a_2^2}$.

In the spin 1/2 the unit ket vectors in the spin 1/2 case are determined by the 2-dimensional Bloch sphere. Hence the following result is intuitive from Corollary 5.2.

Proposition 5.4 *For the spin 1/2 case the vectors of Proposition 5.3 give all unit vectors in the Hilbert space of dimension 2.*

Proof Let v_0 be a fixed 2-dimensional complex unit vector, e.g., $v_0 = (1, 0)$, and let v be any complex vector of dimension 2. Then there is a unitary matrix M with determinant 1 such that $v = Mv_0$. It is well known (see, e.g., Ma 2007) that there is a homomorphism of the group $SU(2)$ of unitary matrices with determinant 1 onto the 3-dimensional rotation group. Let R be the image of M under this homomorphism. Fix a fixed direction a_0, let $a = Ra_0$, and look at the ket vector $|a; +\rangle$. This gives a mapping from v to $|a; +\rangle$. Changing a into $-a$ in this argument gives a mapping from the complex unit 2-vectors to the ket vectors $|a; -\rangle$. By Proposition 5.3 each ket vector $|a; k\rangle$ is a complex unit vector. Hence we have established a one-to-one correspondence.

For $j > 1/2$ the set of simple epistemic states for angular momenta is not closed under linear combinations. As remarked in Sect. 1.5.7, however, each pure quantum state can nevertheless be seen as the eigenstate of some operator; the problem is only to associate this operator to a physically meaningful e-variable. For $j > 1/2$ it is not enough to look at angular momentum components θ^a in various directions a. Candidates for other physically meaningful e-variables may be, e.g., $\alpha\theta^a + \beta\theta^b$ for $a \neq b$, or more generally any $f(\theta^a, \theta^b, \theta^c, \ldots)$.

5.3 Continuous e-Variables: Phase Space

Consider the one-dimensional movement of a single non-relativistic particle in some force field, the particle having position ξ and momentum π at some given time. Both ξ and π are e-variables and can be estimated by suitable experiments. But it has been well known from the early days of quantum mechanics that it is impossible to estimate the vector $\phi = (\xi, \pi)$ with arbitrary accuracy. Thus the point ϕ in the phase space is an inaccessible conceptual variable.

I will first concentrate in the position ξ. This is a continuous variable, so a state cannot be defined as simply as in the discrete case. Consider a fixed confidence interval or credibility interval $(\underline{\xi}, \bar{\xi})$ for this position. Either ξ lies in this interval or it does not lie in this interval. In the first case, the confidence coefficient/credibility coefficient of the interval can be made arbitrarily close to 1 by doing a suitable large experiment. In the second case, the same coefficient can be made arbitrarily close to 0. Thus it is crucial by experiment, that is, by an epistemic process, to make a choice between the two indicator variables:

$$I_1(\xi) = I(\xi \in (\underline{\xi}, \bar{\xi})), \quad I_2(\xi) = I(\xi \notin (\underline{\xi}, \bar{\xi})).$$

Let G^ξ be the translation group on the real line $\mathcal{R}$. The invariant measure corresponding to G^ξ is the Lebesgue measure $d\xi$, and I will define the Hilbert space $H = L^2(\mathcal{R}, d\xi)$. The indicator $I_1(\xi)$ belongs to this space. The indicator $I_2(\xi)$ does not belong to H, but this is not important since $I_2 = 1 - I_1$.

By letting $\underline{\xi}$ and $\bar{\xi}$ vary, the ξ-state of the system can be defined in terms of the indicators I_1. For fixed $\underline{\xi}$ and $\bar{\xi}$ this is a discrete e-variable taking values $I_1 = 0$ and $I_1 = 1$. It is crucial in quantum mechanics that linear combinations of states defined by indicators and the limits of these also can be introduced as states. One way they will emerge is through the time development of states through the Schrödinger equation; see Sect. 5.11.

The approach I will take here is a limiting operation obtained through dividing the real line into many intervals such that the width of each interval tends to zero. Through this limiting process we can approximate any function f in H. In traditional quantum mechanics, any such f is describing a state of the particle, and f is called a wave function. I will not go into any interpretation of this here, but just mention that there are interpretations trying to connect this to the theory of stochastic processes; this is the content of the stochastic mechanics of Edward Nelson (1967). I will discuss this later in connection to the Schrödinger equation, but here I only address the limiting process. I will limit myself to continuous f.

Thus for each n let $\xi_{n1} < \xi_{n2} < \ldots < \xi_{nk_n}$ be a sequence of real numbers such that

1. $\xi_{n1} \to -\infty$ and $\xi_{nk_n} \to \infty$ as $n \to \infty$.
2. $\delta_n = \xi_{n,i+1} - \xi_{ni}$ is constant for $i = 1, \ldots, k_n - 1$ and tends to 0 as $n \to \infty$.

Let $I_{ni}(\xi) = I(\xi \in (\xi_{ni}, \xi_{n,i+1}])$ for $i = 1, 2, \ldots, k_n - 1$. For a given continuous function $f \in H$, define the step function approximation f_n by $f_n(\xi) = f(\xi_{ni})$ for $\xi_{ni} \le \xi < \xi_{n,i+1}$ when $i = 1, \ldots, k_n - 1$; $f_n(\xi) = 0$ for $\xi < \xi_{n1}$ and for $\xi \ge \xi_{nk_n}$. Thus $f_n(\xi) = \sum_i f(\xi_{ni}) I_{ni}(\xi)$, a linear function of indicators. Finally, on the space of such step functions define the operator A_n by

$$A_n f_n(\xi) = \sum_i \xi_{ni} f(\xi_{ni}) I_{ni}(\xi). \tag{5.7}$$

The interpretation of (5.7) is as follows: Approximate ξ by ξ_{ni} when $\xi \in (\xi_{ni}, \xi_{n.i+1}]$ (and neglect its value when $\xi < \xi_{n1}$ or $\xi \geq \xi_{nk_n}$; this is assumed to have negligible probability/ confidence coefficient). This approximate variable is discrete, so we can use the theory of Sect. 5.1. The indicators I_{ni} can be regarded as an orthonormal set of ket vectors for this approximate variable for a suitable normalization of the Lebesgue measure. Then (5.7) is equivalent to (5.2) of Sect. 5.1 with $|i\rangle = I_{ni}(\xi)$ constituting an orthonormal basis for a Hilbert space H_n of step functions, a subspace of H. Thus A_n is the quantum-mechanical operator of the discrete e-variable.

It is of interest to see what happens when n tends to ∞. The following basic result is proved in Appendix C.

Theorem 5.3 *Assume that f is a continuous function in H such that the function k defined by $k(\xi) = \xi f(\xi)$ satisfies $\|k\| < \infty$. Then $\|f_n - f\| \to 0$ and $\|A_n f_n - k\| \to 0$ as $n \to \infty$.*

In this specific sense the operator A corresponding to the e-variable ξ can be said to be the operator of multiplying with ξ. By Theorem 5.3 it is motivated as such an operator defined on all continuous f in H such that $\int |\xi f(\xi)|^2 d\xi < \infty$.

The operator A is an unbounded operator, and as such it must always have a limited domain of definition $D(A)$. There is a very large and advanced mathematical theory on unbounded operators; see for instance Murphy (1990) or Bing-Ren (1992).

The spectral theorem (see any mathematics book or for instance Busch et al. 1991, page 21) is valid also for the operator A. The spectral measure for an interval $I = |\xi_1, \xi_2]$ is given by $E(I)f(\xi) = f(\xi)$ for $\xi_1 \leq \xi \leq \xi_2$ and $E(I) = 0$ for ξ outside the interval. This $E(I)$ is the operator of multiplying with the indicator function of the interval. This measure will be important when considering general versions of the Born formula; see Sect. 5.7 and Sect. 5.15.

So far I have considered the position ξ. A completely parallel discussion can be made on the moment π in the Hilbert space $H^\pi = L^2(\mathcal{S}, d\pi)$, where $\mathcal{S}$ is the line where π varies. Thus the operator B corresponding to momentum π in this space is multiplication by π with domain of definition $D(B) = \{f \in H^\pi : \int |\pi f(\pi)|^2 d\pi < \infty$.

As in the discrete case it is important to have everything described in one Hilbert space, so we need a unitary transformation from H^π to H. For this case we have an apparently different solution than I offered in the maximal symmetric epistemic setting, namely the use of Fourier transform. If $\widehat{f} \in H^\pi$, we define the corresponding $f \in H$ by

$$(U\widehat{f})(\xi) = f(\xi) = \frac{1}{\sqrt{2 \cdot 3.14}} \int \exp(i\frac{\xi\pi}{\hbar})\widehat{f}(\pi)d\pi,$$

where $\hbar$ is Planck's constant, which has the correct unit of measurement. One point here is that this unitary transformation does not transform indicator variables into indicator variables, so there is no confusion between simple π-states and simple

ξ-states. The inverse transformation is given by

$$(U^\dagger f)(\pi) = \widehat{f}(\pi) = \frac{1}{\sqrt{2 \cdot 3.14}} \int \exp(-i\frac{\xi\pi}{\hbar}) f(\xi) d\xi.$$

These two equations may be considered as the continuous analogues of (5.3).

By partial integration one can show that the operator $C = UBU^\dagger$ corresponding to B in H is given by $-i\hbar\frac{d}{d\xi}$ with domain of definition $D(C)$ given by the set of differentiable f such that $\int |f'(\xi)|^2 d\xi < \infty$. It follows that when $f \in D(A) \cap D(C)$ we have $(AC - CA)f(\xi) = i\hbar f(\xi)$, so A and C do not commute. Hence by the brief discussion of commutation relations in Sect. 5.1, ξ and π cannot be estimated simultaneously with arbitrary accuracy, in agreement with observed fact. From the commutation relation Heisenberg's uncertainty relation can be proved: From any estimators $\widehat{\xi}$ and $\widehat{\pi}$ we have $\mathrm{std}(\widehat{\xi})\mathrm{std}(\widehat{\pi}) > \hbar/2$. For a derivation, see standard quantum-mechanical texts or Holevo (2001). But now I am anticipating the inference theory which will be briefly discussed in later sections.

5.4   A Link to Statistical Inference

In this section I again assume the discrete quantum formulation of Sect. 5.1. We can here think of a spin component in a fixed direction to be determined.

Assume now that the quantum mechanical system is prepared in a state with an unknown value of θ^a. Given then the focused question a, the e-variable θ^a plays the role similar to a parameter in statistical inference, even though it may be connected to a single unit. Inference can be done by preparing many independent units in the same state. Inference is then from data z^a, a part of the total data z that nature can provide us with. All inference theory that one finds in standard texts like Lehmann and Casella (1998) applies. In particular, the concepts of unbiasedness, equivariance, average risk optimality, minimaxity and admissibility apply. None of these concepts are much discussed in the physical literature, first because measurements there are often considered as perfect, at least in elementary texts, secondly because, when measurements are considered in the physical literature, they are discussed in terms of the more abstract concept of an operator-valued measure; see Sect. 5.7 below.

Whatever kind of inference we make on θ^a, we can take as a point of departure the statistical model and the generalized likelihood principle of Sect. 3.2.4. Hence after an experiment is done, and given some context τ, all evidence on θ^a is contained in the likelihood $p(z^a|\tau, \theta^a)$, where z^a is the portion of the data relevant for inference on θ^a, also assumed discrete. This is summarized in the likelihood effect:

$$E^a(\boldsymbol{u}^a; z^a, \tau) = \sum_k p(z^a|\tau, \theta^a = u_k^a)|a; k\rangle\langle a; k|.$$

Interpretation of the Likelihood Effect $E^a(z^a, \tau)$ *(1) We have posed some inference question on the accessible e-variable θ^a. (2) We have specified the relevant likelihood for the data. The likelihood for all possible answers of the question, formulated in terms of state vectors, can be recovered from the likelihood effect.*

Since the focused question assumes discrete data, each likelihood is in the range $0 \le p \le 1$. In the quantum mechanical literature, an effect is any operator with eigenvalues in the range $[0, 1]$.

Some qualifications must be made relative to this interpretation, however, if we want to be precise. We have the freedom to redefine the e-variable in the case of coinciding eigenvalues in the likelihood effect, that is, if $p(z^a | \tau, \theta^a = u_k^a) = p(z^a | \tau, \theta^a = u_l^a)$ for some k, l. An extreme case is the likelihood effect $E(u^a; z^a, \tau) = I$, where all the likelihoods are 1, that is, the probability of z is 1 under any considered model.

Recall that the e-variable θ^a is identified with the Hilbert space operator $\sum_k u_k^a \Pi_K^a$, cf. (4.1). Hence the projections $|\Pi_k^a k|$ are crucial in this identification.

We have the following mathematical result on the likelihood effects:

Proposition 5.5 *Assume that a and b are such that there exist some constant $c > 0$ such that*

$$E^a(u^a; z^a, \tau) = c E^b(u^b; z^b, \tau). \tag{5.8}$$

Then we can order the states such that

1) $p(z^a | \tau, \theta^a = u_k^a) = cp(z^b | \tau, \theta^b = u_k^b)$ for each k.
2) Introduce the class of indices C_j such that $p(z^a | \tau, \theta^a = u_k^a) = p(z^a | \tau, \theta^a = u_l^a)$ whenever $k, l \in C_j$, and these likelihoods are different when k and l belong to different C_j-classes, similarly D_j for $p(z^b | \tau, \theta^b = u_k^b)$. Then we have

$$\sum_{k \in C_j} |a; k\rangle\langle a; k| = \sum_{k \in D_j} |b; k\rangle\langle b; k|$$

for all j.

On the other hand, if 1) and 2) are satisfied, then (5.8) holds.

The last part is fairly trivial. The direct part is proved in Appendix C.

Return now to the generalized likelihood principle of Sect. 3.2.4. Recall that this principle follows from the very reasonable conditionality principle and sufficiency principle, and that the principle itself is fairly reasonable in our setting, where we condition upon the context τ. In statistics, the likelihood principle says the following: If two experiments have proportional likelihood, with constant of proportionality independent of the parameter, they produce the same experimental evidence about the parameter. Here experimental evidence is left undefined. In quantum mechanics, where we focus on a specific question a, we must in addition

demand that this focused question is the same for each possible value of the
e-variable; that is, the projections Π_k^a must be the same.

An observation z^a is called *regular* if $p(z^a|\tau, \theta^a = u_k^a) = p(z^a|\tau, \theta^a = u_j^a)$
implies $u_k^a = u_j^a$.

The following principle follows:

The Focused Generalized Likelihood Principle (FGLP) *Consider two potential
experiments a and b in some setting with equivalent contexts τ, and assume that the
inaccessible conceptual variable ϕ is the same in both experiments. Suppose that
the two observations z_1^a and z_2^b have equal likelihood effects in the two experiments,
and assume that both observations z_1^a and z_2^b are regular.*
Then

A) *The questions posed in the two experiments are equivalent in the sense that the
   corresponding eigenvector spaces are equal.*
B) *The two observations produce equivalent experimental evidence on the relevant
   e-variables in this context.*

Proposition 5.6 *The focused generalized likelihood principle follows from the
generalized likelihood principle of Sect. 3.2.4.*

Proof The Proposition is trivial if we know that the e-variables are the same in the
two experiments. If not, we have a siuation where (Eq. 5.8) holds with $c = 1$. Then
by (Eq. 4.1) θ^a-operator $= \sum u_k^a \Pi_k^a$, θ^b-operator $= \sum v_k^b \Pi_k^a$, where the equality
of the projection operators follow from Proposition 5.5, 2) and the assumption of
regularity. The eigenvalues u_k^a are all different, similarly the eigenvalues v_k^b. These
are the answers to the questions connected to θ^a and θ^b; the questions are equivalent.
B) follows from Proposition 5.5, 1) with $c = 1$ and the ordinary likelihood principle.

Two contexts are considered equivalent if they are one-to-one functions of each
other. For regular observations the principle says that both the question posed and
the experimental evidence are functions of the likelihood effect and the context of
the experiment.

5.5   Rationality and Experimental Evidence

Throughout this section I will consider a fixed context τ and a fixed epistemic
setting in this context. The inaccessible conceptual variable is ϕ, and I assume that
the accessible e-variables θ^a take a discrete set of values. Let the data behind the
potential experiments be z^a, also assumed to take a discrete set of values.

Let first a single experimentalist A be in this situation, and let all conceptual
variables be attached to A, although he also has the possibility to receiving
information from others through part of the context τ. He has the choice of doing
different experiments a, and he also has the choice of choosing different models
for his experiment through his likelihood $p_A(z^a|\tau, \theta^a)$. The experiment and the

model, hence the likelihood, should be chosen before the data are obtained. All these choices are summarized in the likelihood effect E^a, a function of the at present unknown data z^a. For use after the experiment, he should also choose a good estimator $\widehat{\theta^a}$, and he may also have to choose some loss function, but the principles behind these latter choices will be considered as part of the context τ.

If A chooses to do a Bayesian analysis, the estimator should be based on a prior $\pi_A(\theta^a|\tau)$. We assume that he is trying to be as rational as possible in all his choices, and that this rationality is connected to his loss function or to other criteria. What should be meant by experimental evidence, and how should it be measured? As a natural choice, let the experimental evidence that we are seeking, be the posterior probability for some fixed value of θ^a, given the data. From the experimentalist A's point of view this is given by:

$$p_A^a(\theta^a = u_j^a|z^a, \tau) = \frac{p_A(z^a|\tau, \theta^a = u_j^a)\pi_A(\theta^a = u_j^a|\tau)}{p_A^a(z^a|\tau)},$$

assuming the likelihood chosen by A and A's prior π_A for θ^a.

Some Bayesians claim that their own philosophy is the only one which is consistent with the likelihood principle. For my own view on this, see below and also comments in Sect. 3.4 and at the end of Sect. 2.1.3. In a non-Bayesian analysis, we can let the concept of experimental evidence be tied to the confidence distribution, given the context, see Sect. 2.1.3. Also in such an analysis we must assume A to be as rational as possible.

We have to make precise in some way what is meant by the rationality of the experimentalist A. He has to make many difficult choices on the basis of uncertain knowledge. His actions can partly be based on intuition, partly on experience from similar situations, partly on a common scientific culture and partly on advices from other persons. These other persons will in turn have their intuition, their experience and their scientific education. Often A will have certain explicitly formulated principles on which to base his decisions, but sometimes he has to dispense with the principles. In the latter case, he has to rely on some 'inner voice', a conviction which tells him what to do.

We will formalize all this by introducing a perfectly rational superior actor D, to which all these principles, experiences and convictions can be related. We also assume that D can observe everything that is going on, in particular A, and that he on this background can have some influence on A's decisions. We assume that D has priors, so that he can do a Bayesian analysis. These priors can for example be based on symmetry. The real experimental evidence will then be defined as *the probability of the e-variable θ^a from D's point of view, which we assume also to give the real objective probabilities*. By the FGLP this must again be a function of the likelihood effect E^a.

$$D\text{'s experimental evidence} = \text{real probability} = q(E^a(\boldsymbol{u}^a; z^a, \tau)) \qquad (5.9)$$

under the assumption that the experiment connected to some e-variable θ^a is to be done. Here and in the following I will assume that the observations z^a are regular in the sense defined in Sect. 5.4 . I will come back to this assumption later.

The superior actor D represents here the scientific ideals of the experimentalist A, and my main point is that D should be perfectly rational. This should influence A's decisions during the experiment.

In this book I have not tried to develop a theory of decisions. Instead I have in Sect. 1.3 referred to the Quantum Decision Theory of Yukalov and Sornette, which is particularly nicely formulated in Yukalov and Sornette (2010) and Yukalov and Sornette (2014). Here one must be careful, however. Quantum Decision Theory takes its departure in the ordinary mathematical formulation of quantum theory, in particular in Born's rule for calculating probabilities. Using this theory here, where I am preparing to derive Born's rule, will lead to circular reasoning. Instead I will now take *decision* as a primitive concept.

An important point is that decisions made by the brain are not the same as straightforward computerlike calculations. Human decisions are based on the functioning of and the interplay between conscious and subconscious processes in the brain. Informally one might say that D also contains an idealization of what makes the mind of A different from a computer. In simple words, this might be called intuition. Intuition is the source of energy for all sorts of art, for communication between people, and even for scientific investigations.

In a scientific connection we assume that D is perfectly rational. This can be formalized mathematically by considering a hypothetical betting situation for D against a bookie, nature N. A similar discussion was recently done using a more abstract language by Hammond (2011). Note the difference to the ordinary Bayesian assumption, where A himself is assumed to be perfectly rational. This difference is crucial to me. I do not see any human scientist, including myself, as being perfectly rational. We can try to be as rational as possible, but we have to rely on some underlying rational ideals that partly determine our actions.

So let the hypothetical odds of a given bet for D be $(1 - q)/q$ to 1, where q is the probability as defined by (5.9). This odds specification is a way to make precise that, given the context τ and given the question a, the bettor's probability that the experimental result takes some value is given by q: For a given utility measured by x, the bettor D pays in an amount qx—the stake—to the bookie. After the experiment the bookie pays out an amount x—the payoff—to the bettor if the result of the experiment takes the value θ^a, otherwise nothing is payed.

The rationality of D is formulated in terms of

The Dutch Book Principle *No choice of payoffs in a series of bets shall lead to a sure loss for the bettor.*

For a related use of the same principle, see Caves et al. (2002).

Assumption 5.5 *Consider in some context τ an epistemic setting where the FGLP is satisfied, and the whole situation is observed and acted upon by a superior actor D as described above. Assume that D's probabilities q given by (5.9) are taken as*

*the experimental evidence, and that D acts rationally in agreement with the Dutch
book principle.*

A situation where Assumptions 5.4 and 5.5 hold will be called a *rational
epistemic setting*. It will be assumed to be implied by essential situations of quantum
mechanics. The question will be raised later if it also can be coupled to certain
macroscopic situations.

Theorem 5.4 *Assume a rational epistemic setting, and assume a fixed context τ.
Let E_1 and E_2 be two likelihood effects in this setting, and assume that $E_1 + E_2$
also is an effect. Then the experimental evidences, taken as the probabilities of the
corresponding data, satisfy*

$$q(E_1 + E_2|\tau) = q(E_1|\tau) + q(E_2|\tau).$$

Proof The result of the theorem is obvious, without making Assumption 5.5, if E_1
and E_2 are likelihood effects connected to experiments on the same e-variable θ^a.
We will prove it in general. Consider then any finite number of potential experiments
including the two with likelihood effects E_1 and E_2. Let $q_1 = q(E_1|\tau)$ be equal to
(5.9) for the first experiment, and let $q_2 = q(E_2|\tau)$ be equal to the same quantity for
the second experiment. Consider in addition the following randomized experiment:
Throw an unbiased coin. If head, choose the experiment with likelihood effect E_1; if
tail, choose the experiment with likelihood effect E_2. This is a valid experiment. The
likelihood effect when the coin shows head is $\frac{1}{2}E_1$, when it shows tail $\frac{1}{2}E_2$, so that
the likelihood effect of this experiment is $E_0 = \frac{1}{2}(E_1 + E_2)$. Define $q_0 = q(E_0)$. Let
the bettor bet on the results of all these three experiments: Payoff x_1 for experiment
1, payoff x_2 for experiment 2 and payoff x_0 for experiment 0.

I will divide into three possible outcomes: Either the likelihood effect from the
data z is E_1 or it is E_2 or it is none of these. The randomization in the choice of E_0
is considered separately from the result of the bet. (Technically this can be done by
repeating the whole series of experiments many times with the same randomization.
This is also consistent with the conditionality principle.) Thus if E_1 occurs, the
payoff for experiment 0 is replaced by the expected payoff $x_0/2$, similarly if E_2
occurs. The net expected amount the bettor receives is then

$$x_1 + \frac{1}{2}x_0 - q_1x_1 - q_2x_2 - q_0x_0 = (1 - q_1)x_1 - q_2x_2 - (1 - 2q_0)\frac{1}{2}x_0 \text{ if } E_1,$$

$$x_2 + \frac{1}{2}x_0 - q_1x_1 - q_2x_2 - q_0x_0 = -q_1x_1 - (1 - q_2)x_2 - (1 - 2q_0)\frac{1}{2}x_0 \text{ if } E_2,$$

$$-q_1x_1 - q_2x_2 - 2q_0 \cdot \frac{1}{2}x_0 \text{ otherwise.}$$

The payoffs (x_1, x_2, x_0) can be chosen by nature N in such a way that it leads to sure loss for the bettor D if not the determinant of this system is zero:

$$0 = \begin{vmatrix} 1-q_1 & -q_2 & 1-2q_0 \\ -q_1 & 1-q_2 & 1-2q_0 \\ -q_1 & -q_2 & -2q_0 \end{vmatrix} = q_1 + q_2 - 2q_0.$$

Thus we must have

$$q(\frac{1}{2}(E_1 + E_2)|\tau) = \frac{1}{2}(q(E_1|\tau) + q(E_2|\tau)).$$

If $E_1 + E_2$ is an effect, the common factor $\frac{1}{2}$ can be removed by changing the likelihoods, and the result follows.

Corollary *Assume a rational epistemic setting in the context τ. Let E_1, E_2, ... be likelihood effects in this setting, and assume that $E_1 + E_2 + \ldots$ also is an effect. Then*

$$q(E_1 + E_2 + \ldots |\tau) = q(E_1|\tau) + q(E_2|\tau) + \ldots.$$

Proof The finite case follows immediately from Theorem 5.4. Then the infinite case follows from monotone convergence.

The result of this section is quite general. In particular the loss function and any other criterion for the success of the experiments are arbitrary. So far I have assumed that the choice of experiment a is given, which implies that it is the same for A and for D. However, the result also applies to the following different situation: Let A have some definite purpose for his experiment, and to achieve that purpose, he has to choose the question a in a clever manner, as rationally as he can. Assume that this rationality is formalized through the actor D, who has the ideal likelihood effect E and the experimental evidence $q(E|\tau)$. If two such questions can be chosen, the result of Theorem 5.4 holds, with essentially the same proof.

5.6   The Born Formula

5.6.1   The Basic Formula

Born's formula is the basis for all probability calculations in quantum mechanics. In textbooks it is usually stated as a separate axiom, but it has also been argued for by using various sets of assumptions; see Helland (2008) for some references. In fact, the first argument for the Born formula, assuming that there is an affine mapping from set of density functions to the corresponding probability functions, is due to von Neumann (1927) (see Busch et al. 2016, p. 201). In Helland (2006,

2008, 2010) the formula was proved under rather strong assumptions. Here I will use assumptions which are as weak as possible; I will base the discussion upon the result of Sect. 5.5.

I begin with a very elegant recent theorem by Busch (2003). For completeness I reproduce the proof for the finite-dimensional case in Appendix D.

Let in general H be any separable Hilbert space. Recall that an effect E is any operator on the Hilbert space with eigenvalues in the range $[0, 1]$. A generalized probability measure μ is a function on the effects with the properties

(1) $0 \le \mu(E) \le 1$ for all E,
(2) $\mu(I) = 1$,
(3) $\mu(E_1 + E_2 + \ldots) = \mu(E_1) + \mu(E_2) + \ldots$ whenever $E_1 + E_2 + \ldots \le I$.

Theorem 5.5 (Busch 2003) *Any generalized probability measure μ is of the form $\mu(E) = \text{trace}(\rho E)$ for some density operator ρ.*

It is now easy to see that $q(E|\tau)$ on the ideal likelihood effects of Sect. 5.5 is a generalized probability measure if Assumption 5.5 holds: (1) follows since q is a probability; (2) since $E = I$ implies that the likelihood is 1 for all values of the e-variable; finally (3) is a consequence of the corollary of Theorem 5.4. Hence there is a density operator $\rho = \rho(\tau)$ such that $p(z|\tau) = \text{trace}(\rho(\tau)E)$ for all ideal likelihood effects $E = E(z)$. This is a result which is valid for all experiments.

The problem of defining a generalized probability on the set of effects is also discussed in Busch et al. (2016).

Define now a *perfect experiment* as one where the measurement uncertainty can be disregarded. The quantum mechanical literature operates very much with perfect experiments which give well-defined states $|j\rangle$. From the point of view of statistics, if, say the 99% confidence or credibility region of θ^b is the single point u_j^b, we can infer approximately that a perfect experiment has given the result $\theta^b = u_j^b$.

For perfect experiments the observations z^a are essentially equal to the relevant e-variable values $\theta^a = u_k^a$, so we can without loss of generality assume that these observations are regular, (see Sect. 5.4).

In our epistemic setting then: We have asked the question: 'What is the value of the accessible e-variable θ^b?', and are interested in finding the probability of the answer $\theta^b = u_j^b$ though a perfect experiment. If u_j^b is a non-degenerate eigenvalue of the operator corresponding to θ^b, this is the probability of a well-defined state $|b; j\rangle$. Assume now that this probability is sought in a context $\tau = \tau^{a,k}$ defined as follows: We have previous knowledge of the answer $\theta^a = u_k^a$ of another minimal question: 'What is the value of θ^a?' That is, we know the state $|a; k\rangle$.

Theorem 5.6 (Born's formula) *Assume a rational epistemic setting. In the above situation we have:*

$$P(\theta^b = u_j^b | \theta^a = u_k^a) = |\langle a; k|b; j\rangle|^2.$$

Proof Assume that both the e-variable θ^a and the e-variable θ^b have operators with non-degenerate eigenvalues. Fix j and k, let $|v\rangle$ be either $|a; k\rangle$ or $|b; j\rangle$, and consider likelihood effects of the form $E = |v\rangle\langle v|$. This corresponds in both cases to a perfect measurement of a maximally accessible parameter with a definite result. By Theorem 5.5 there exists a density operator $\rho^{a,k} = \sum_i \pi_i(\tau^{a,k})|i\rangle\langle i|$ such that $q(E|\tau^{a,k}) = \langle v|\rho^{a,k}|v\rangle$, where $\pi_i(\tau^{a,k})$ are non-negative constants adding to 1. Consider first $|v\rangle = |a; k\rangle$. For this case one must have $\sum_i \pi_i(\tau^{a,k})|\langle i|a; k\rangle|^2 = 1$ and thus $\sum_i \pi_i(\tau^{a,k})(1 - |\langle i|a; k\rangle|^2) = 0$. This implies for each i that either $\pi_i(\tau^{a,k}) = 0$ or $|\langle i|a; k\rangle| = 1$. Since the last condition implies $|i\rangle = |a; k\rangle$ (modulus an irrelevant phase factor), and this is a condition which can only be true for one i, it follows that $\pi_i(\tau^{a,k}) = 0$ for all other i than this one, and that $\pi_i(\tau^{a,k}) = 1$ for this particular i. Summarizing this, we get $\rho^{a,k} = |a; k\rangle\langle a; k|$, and setting $|v\rangle = |b; j\rangle$, Born's formula follows, since $q(E|\tau^{a,k})$ in this case is equal to the probability of the perfect result $\theta^b = u_j^b$.

5.6.2   Consequences

Here are three easy consequences of Born's formula:

1. If the context of the system is given by the state $|a; k\rangle$, and A^b is the operator corresponding to the e-variable θ^b, then the expected value of a perfect measurement of θ^b is $\langle a; k|A^b|a; k\rangle$.
2. If the context is given by a density operator ρ, and A is the operator corresponding to the e-variable θ, then the expected value of a perfect measurement of θ is trace(ρA). This is the formula used in Postulate 1.2 of Sect. 1.5.5.
3. In the same situation the expected value of a perfect measurement of $f(\theta)$ is trace($\rho f(A)$).

Proof of (1)

$$\mathrm{E}(\theta^b|\theta^a = u_k^a) = \sum_i u_i^b P(\theta^b = u_i^b|\theta^a = u_k^a)$$

$$= \sum_i u_i^b \langle a; k|b; i\rangle\langle b; i|a; k\rangle = \langle a; k|A^b|a; k\rangle.$$

Proof of (2) Let $\rho = \sum_k \pi_k^a|a; k\rangle\langle a; k|$ and $A = \sum_j u_j^b|b; j\rangle\langle b; j|$. Then from 1)

$$\mathrm{E}(\theta) = \sum_k \pi_k^a \langle a; k|A|a; k\rangle = \mathrm{trace} \sum_k \pi_k^a|a; k\rangle\langle a; k|A.$$

These results give an extended interpretation of the operator A compared to what I gave in Sect. 5.4: There is a simple formula for all expectations in terms of the operator. On the other hand, the set of such expectations determine the state of the system. Also on the other hand: If A is specialized to an indicator function, we get back Born's formula, so the consequences are equivalent to this formula.

As an application of Born's formula, we give the transition probabilities for electron spin. Throughout this book, we will, for a given direction a, define the e-variable θ^a as $+1$ if the measured spin component by a perfect measurement for the electron is $+\hbar/2$ in this direction, $\theta^a = -1$ if the component is $-\hbar/2$. Assume that a and b are two directions in which the spin component can be measured.

Proposition 5.7 *For electron spin we have*

$$P(\theta^b = \pm 1 | \theta^a = +1) = \frac{1}{2}(1 \pm \cos(a \cdot b)).$$

This is proved in several textbooks, for instance Holevo (2001), from Born's formula. A similar proof using the Pauli spin matrices is also given in Helland (2010).

Measurements of e-variables will be treated in Sect. 5.7, but here we will look at the simpler case of a perfect measurement. Assume that we know the state $|\psi\rangle$ of a system, and that we want to measure a new e-variable θ^b. This can be discussed by means of the projection operators $\Pi_j^b = |b; j\rangle\langle b; j|$. First observe that by a simple calculation from Born's formula

$$P(\theta^b = u_j^b | \psi) = \|\Pi_j^b |\psi\rangle\|^2. \tag{5.10}$$

Next observe that after a perfect measurement $\theta^b = u_j^b$ have been obtained, the state changes to

$$|b; j\rangle = \frac{\Pi_j^b |\psi\rangle}{\|\Pi_j^b |\psi\rangle\|}.$$

Successive measurements are often of interest. We find

$$P(\theta^b = u_j^b \text{ and then } \theta^c = u_i^c | \psi) = P(\theta^c = u_i^c | \theta^b = u_j^b) P(\theta^b = u_j^b | \psi)$$

$$= \|\Pi_i^c \frac{\Pi_j^b |\psi\rangle}{\|\Pi_j^b |\psi\rangle\|}\|^2 \|\Pi_j^b |\psi\rangle\|^2 = \|\Pi_i^c \Pi_j^b |\psi\rangle\|^2. \tag{5.11}$$

In the case with multiple eigenvalues, the formulae above are still valid, but the projectors above must be replaced by projectors upon eigenspaces. One can show that (5.10) then gives a precise version of Born's rule for this case.

Proof Look first at the case with unique eigenvalues. Then Born's rule says

$$P(\theta^b = u_j^b | \psi) = \langle \psi | b; j\rangle\langle b; j | \psi\rangle.$$

Let then the eigenvalues move towards coincidence. Let $C_k = \{j : u_j^b = v_k^b\}$. Then by continuity from the previous equation we get

$$P(\theta^b = v_k^b | \psi) = \sum_{j \in C_k} \langle \psi | b; j \rangle \langle b; j | \psi \rangle = \langle \psi | \Pi_k^b | \psi \rangle = \| \Pi_k^b | \psi \rangle \|^2.$$

Note that in general $P(\theta^b = u_j^b$ and then $\theta^c = u_i^c | \psi) \neq P(\theta^c = u_i^c$ and then $\theta^b = u_j^b | \psi)$. Measurements do not necessarily commute.

5.6.3 A Macroscopic Example

A very relevant question is now: Are all these results, including Born's formula, by necessity confined to the microworld? Recently, physicists have become interested in larger systems where quantum mechanics is valid, see Vedral (2011). Of even more interest are the quantum models of cognition, see Pothos and Busemeyer (2013). As we have defined it, there is nothing microscopic about the epistemic setting. It may or may not be that the rationality Assumption 5.5 also is valid for some larger scale systems. The following example illustrates the point.

Example 5.16 In a medical experiment, let μ_a, μ_b, μ_c and μ_d be continuous inaccessible parameters, the hypothetical effects of treatment a, b, c and d, respectively. Assume that the focus of the experiment is to compare treatment b with the mean effect of the other treatments, which is supposed to give the parameter $\frac{1}{3}(\mu_a + \mu_c + \mu_d)$. One wants to do a pairwise experiment, but it turns out that the maximal parameter which can be estimated, is

$$\theta^b = \text{sign}(\mu^b - \frac{1}{3}(\mu_a + \mu_c + \mu_d)).$$

(Imagine for example that one has four different ointments against rash. A patient is treated with ointment b on one side of his back; a mixture of the other ointments on the other side of his back. It is only possible to observe which side improves best, but this observation is assumed to be very accurate. One can in principle do the experiment on several patients, and select out the patients where the difference is clear.) This experiment is done on a selected set of experimental units, on whom it is known from earlier accurate experiments that the corresponding parameter

$$\theta^a = \text{sign}(\mu^a - \frac{1}{3}(\mu_b + \mu_c + \mu_d))$$

takes the value $+1$. In other words, one is interested in the probabilities

$$\pi = P(\theta^b = +1 | \theta^a = +1).$$

Consider first a Bayesian approach. Natural priors for $\mu_a, \ldots, \mu_d$ are independent $N(v, \sigma^2)$ with the same v and σ. By location and scale invariance, there is no loss in generality by assuming $v = 0$ and $\sigma = 1$. Then the joint prior of $\zeta^a = \mu_a - \frac{1}{3}(\mu_b + \mu_c + \mu_d)$ and $\zeta^b = \mu_b - \frac{1}{3}(\mu_a + \mu_c + \mu_d)$ is multinormal with mean $\mathbf{0}$ and covariance matrix

$$\begin{pmatrix} \frac{4}{3} & -\frac{4}{9} \\ -\frac{4}{9} & \frac{4}{3} \end{pmatrix}.$$

A numerical calculation from this gives

$$\pi = P(\zeta^b > 0 | \zeta^a > 0) \approx 0.43.$$

This result can also be assumed to be valid when $\sigma \to \infty$, a case which in some sense can be considered as independent objective priors for $\mu_a, \ldots, \mu_d$.

Now consider a rational epistemic setting for this experiment. Since again scale is irrelevant, a natural group on $\mu_a, \ldots, \mu_d$ is a 4-dimensional rotation group around a point $(v, \ldots, v)$ together with a translation of v. Furthermore, ζ^a and ζ^b are contrasts, that is, linear combinations with coefficients adding to 0. The space of such contrasts is a 3-dimensional subspace of the original 4-dimensional space, and by a single orthogonal transformation, the relevant subset of the 4-dimensional rotations can be transformed into the group G of 3-dimensional rotations on this latter space, and the translation in v is irrelevant. One such orthogonal transformation is given by

$$\psi_0 = \frac{1}{2}(\mu_a + \mu_b + \mu_c + \mu_d),$$

$$\psi_1 = \frac{1}{2}(-\mu_a - \mu_b + \mu_c + \mu_d),$$

$$\psi_2 = \frac{1}{2}(-\mu_a + \mu_b - \mu_c + \mu_d),$$

$$\psi_3 = \frac{1}{2}(-\mu_a + \mu_b + \mu_c - \mu_d).$$

Let G be the group of rotations orthogonal to ψ_0. We find

$$\zeta^a = -\frac{2}{3}(\psi_1 + \psi_2 + \psi_3),$$

$$\zeta^b = -\frac{2}{3}(\psi_1 - \psi_2 - \psi_3).$$

The rotation group element transforming ζ^a into ζ^b is homomorphic under G to the rotation group element g_{ab} transforming $a = -\frac{1}{\sqrt{3}}(1, 1, 1)$ into $b = -\frac{1}{\sqrt{3}}(1, -1, -1)$. Let G^a be the maximal subgroup of G under which ζ^a is permissible. This is isomorphic with the group of rotations around a together with a reflection in the plane perpendicular to a, but the action on ζ^a is just a reflection. The orbits of this group are given by two-point sets $\{\pm c\}$. In conclusion, the whole situation is completely equivalent to the spin-example of Example 4.15 and satisfies the assumptions of the symmetrical epistemic setting. Making the rationality Assumption 5.5 then implies from Proposition 5.7:

$$\pi = P(\text{sign}(\zeta^b) = +1 | \text{sign}(\zeta^a) = +1) = \frac{1}{2}(1 + a \cdot b) = \frac{1}{3}.$$

To be precise, Example 5.16 satisfies the assumptions of the maximal symmetrical epistemic setting except Assumptions 4.2a) and 4.2b). To have these assumptions satisfied, we must extend the situation:

Example 5.17 Let the situation be as in Example 5.16 with the addition that we have available treatments with hypothetical effects μ_a for $a \in \mathcal{A}$, where the index set $\mathcal{A}$ can be taken to be the 3-dimensional unit sphere.

It is clear that the extension from Example 5.16 to Example 5.17 does not mean anything for the result. Assumption 4.3 must be replaced by the result of Corollary 5.1 of Sect. 5.2. Assumption 5.4 is of no relevance here.

I guess that many statisticians will prefer the Bayesian calculations here for the rational epistemic setting calculations, which some may consider to have a more speculative foundation. But the prior chosen in this example must be considered somewhat arbitrary, and its 'objective' limit may lead to conceptual difficulties. Since experiments of this kind can in principle be done in practice—at least approximately, the question whether the Bayesian solution or the rational epistemic setting solution holds in such cases, must ultimately be seen as an empirical question.

5.6.4 Superselection Rules

Two states $|\psi\rangle$ and $|\theta\rangle$ obey a superselection rule if $\langle \psi | A | \theta \rangle = 0$ for all operators A representing e-variables. This can be the case for instance if the Hilbert space decomposes as $H = H_1 \oplus H_2$, $|\psi\rangle \in H_1$, $|\theta\rangle \in H_2$ and all e-variables act either on H_1 or H_2. In this case the linear combinations $|\eta\rangle = \alpha|\psi\rangle + \beta|\theta\rangle$ have no physical meaning by the following argument:

$$\langle \eta | A | \eta \rangle = \text{trace}(\rho A)$$

for all A, where $\rho = |\alpha|^2 |\psi\rangle\langle\psi| + |\beta|^2 |\theta\rangle\langle\theta|$, so the artificial superposition might as well be replaced by a density matrix. A thorough discussion of superselection rules can be found in Giulini (2009).

5.7 Measurements and Quantum Statistical Inference

5.7.1 Simple Measurements

It is an important task now to depart from the assumption of perfect measurements, and address measurements with real data z. I start by discussing the simplest measurements. I first introduce the concept of operator-valued measure.

Assume a likelihood $p(z^b | \theta^b = u_j^b)$ for the e-variable θ^b, and assume first for simplicity that the operator corresponding to θ^b has non-degenerate eigenvalues u_j^b. Define an operator-valued measure M by $M(\{z^b\}) = \sum_j p(z^b | \theta^b = u_j^b)|b; j\rangle\langle b; j|$. This is just the likelihood effect corresponding to the observation z^b. The context τ and the values u_j^a of θ^b are assumed fixed. The operatorvalued measure for a set C in the sample space is defined by $M(C) = \sum_{z \in C} M(\{z\})$.

These operators, called POVM (positive-operator valued measures) in the literature, satisfy $M(S) = I$ for the whole sample space and are countably additive. Let the current state be given by the question: 'What is the value of θ^a?', and then the probabilities $\pi^a(u_k^a)$ for the different values $\theta^a = u_k^a$. Then, defining $\rho = \sum_k \pi^a(u_k^a)|a; k\rangle\langle a; k|$ we get $P(C) = \text{trace}(\rho M(C))$ for all sets C in the sample space. This is again proved by a straightforward argument from Born's formula. An independent proof of the same formula in a more general setting is given by Holevo (1982) Proposition 1.6.1.

This gives a statistical model. In Sect. 5.13 below I give a discussion on how this model can be used to estimate ρ when this is completely unknown. This is however not the usual physical situation. Ordinarily the initial state $|a; k\rangle$ is known through our preparation of the system, and we ask a new question: What is the value of θ^b?

The statistical approach to this problem is first to model the response of the measurement apparatus as a function of θ^b. An example of this approach is given in Sect. 1.5.5, the Stern-Gerlach experiment. In general, statistical theory can be used to estimate θ^b from the experiment, see Lehmann and Casella (1998). Bayesian estimation can be used, since we have a prior from Born's formula.

Since θ^b is discrete, and the experiment is often very accurate, we may get a very accurate estimate, nearly the real value. In fact, standard quantum mechanical texts often assume that the real value of θ^b is obtained in such measurements.

One approach to the quantum theory of measurement is to assume a joint entangled quantum model for the object systems and the apparatus. As discussed in detail by Busch et al. (1991), this may lead to logical difficulties if the apparatus also is assumed to be a quantum system, a point of view which assumes an ontological interpretation of the wave function. These difficulties have been attempted solved by many authors in many different ways. One approach is the consistent history approach, see Griffiths (2014, 2017a). In Griffiths (2017a) this approach is used to solve several apparent paradoxes of quantum theory, among others one involving a Mach-Zehnder inferometer. From an epistemic point of view this is an example of a general situation where an observer may define θ^1 =path chosen by a particle and θ^1 =phase of the source, and where the vector (θ^1, θ^2) is inaccessible for all observers.

5.7.2 Collapse of the Wave Packet

Related to the measurement is the phenomenon of collapse of the wave packet. Assume first an initial state $|a; k\rangle$, and then a perfect measurement giving the value $\theta^b = u_j^b$. After the measurement the state then changes to $|b; j\rangle$. This discontinuous change of the state has been considered a great problem in the very common general ontological view on quantum mechanics, problems so great that some physicists adhere to a many-worlds interpretation (see Everett 1973) to cope with it. In our epistemic interpretation the collapse represents no problem. A similar 'collapse' occurs in Bayesian statistics once an observation is made.

The situation is similar, but more complicated when a real measurement is made. To cope with this, Barndorff-Nielsen et al. (2003) introduced the notion of an *instrument*. A simple instrument is one where the state is transformed by projecting onto orthogonal subspaces of the Hilbert space, together spanning the whole space. This is also called the Lüders-von Neumann projection postulate, and is similar to the collapse in the perfect measurement above. It is indicated in op. cit. that more general instruments can be formed by combining this with Schrödinger evaluation (see Sect. 5.11 here) and by forming compound systems.

For completeness I will give a brief description of the Lüders-von Neumann projection postulate, following Barndorff-Nielsen et al. (2003). Let us measure a variable connected to a selfadjoint operator A. Let $\{u_1, \ldots, u_d\}$ be its *distinct* eigenvalues. Let $\Pi(u)$ denote the matrix which projects onto the eigenspace corresponding to the eigenvalue u, not necessarily one-dimensional. Thus $A = \sum_u u \Pi(u)$. Define an operator-valued measure by $M(\{u\}) = \Pi(u)$, and also what is called an instrument by $n(u) = \Pi(u)$. When this instrument is applied to a quantum state given by the density matrix $\rho = \sum_k \pi_k |k\rangle\langle k|$, we obtain the outcome u with probability $\text{trace}(\rho \Pi(u)) = \sum_k \pi_k \|\Pi(u)|k\rangle\|^2$. We may compute that the final state is the mixture, according to the posterior probabilities, given that the initial pure state was $|k\rangle$ and given that the measurement outcome was u, a mixture of the pure states with state vectors equal to the normalized projections

$$\Pi(u)|k\rangle / \|\Pi(u)|k\rangle\|.$$

The probabilistic interpretation is that with probability π_k the quantum system started in the pure state with state vector $|k\rangle$. On measuring the observable A, the state vector is projected into one of its eigenspaces, with probabilities equal to the square lengths of the projections. We observe the corresponding eigenvalue. The posterior state is the mixture of these different pure states according to the posterior distribution of the initial state given data u. Explicitly, this gives the state with density matrix

$$\frac{\Pi(u)\rho\Pi(u)}{\text{trace}(\Pi(u)\rho)}.$$

5.7.3 Perfect Observations as Seen by a Single Observer

In Chap. 4 the quantum state was connected to a focused question together with a definite answer to that question. The different questions where labeled by an index a. The corresponding e-variable θ^a was connected to an operator A^a on the Hilbert space of the physical system. Typically, two such operators A^a and A^b will be non-commuting. This means that no single observer can observe both θ^a and θ^b at the same time.

So fix one observer, let θ be a simple e-variable observed by this observer, and let A be the operator connected to θ. By the spectral theorem or by the discussion in Sect. 4.5 we can write $A = \sum_k u_k \Pi_k$, where the u_k's are different and the orthogonal projectors $\{\Pi_k\}$ satisfy

$$\sum_k \Pi_k = I, \quad \Pi_k^{\dagger} = \Pi_k, \quad \Pi_j \Pi_k = \delta_{jk}\Pi_k. \tag{5.12}$$

Let the projector Π_k project upon a vector space V_k. Then (5.12) means that the orthogonal vector spaces V_k span the whole Hilbert space. The statement $|\psi\rangle \in V_k$ means that $|\psi\rangle$ is an eigenvector of A corresponding to the eigenvalue u_k. The interpretation of this as seen by a single observer, is: He has posed a question to the system 'What is the value of θ?' and has obtained a definite answer '$\theta = u_k$'.

For the case of a non-generate eigenvalue u_k with eigenvector $|k\rangle$, we have $\Pi_k = |k\rangle\langle k|$. In general (5.12) is called a projective decomposition of the identity (PDI).

If the state of the system is given by a density matrix ρ, then by the Born rule

$$p_k = P(\theta = u_k) = \text{trace}(\rho \Pi_k). \tag{5.13}$$

The projectors Π_k have eigenvalues 0 and 1, and in this way they resemble indicator functions. In fact, a PDI divides the Hilbert space into mutually exclusive subspaces. By combining these subspaces, we get a Boolean algebra (σ-algebra in statistical terms), and probabilities may be assigned to this algebra. For instance, $\Pi_j + \Pi_k$ is a projector, and $P(\Pi_j \ OR \ \Pi_k) = P(\theta = u_j \ or \ u_k) = \text{trace}(\rho(\Pi_j + \Pi_k))$, and similarly for 3 or more distinct projectors.

Two PDI's $\{\Pi_k\}$ and $\{\Gamma_k\}$ are called compatible if $\Pi_j\Gamma_k = \Gamma_k\Pi_j$ for every j and k. The we have a refined Boolean algebra with elements $\Pi_j\Gamma_k$ for all combinations. If $B = \sum_k v_k\Gamma_k$ is connected to an e-variable γ, then $P(\Pi_j \ AND \ \Gamma_k) = P(\theta = u_j \ or \ \gamma = v_k) = \text{trace}(\rho(\Pi_j\Gamma_k))$. In this way a complete probability calculus for ideal observations may be found. In the literature on consistent histories, see for instance Griffiths (2017b), it is important that $P(\Pi_j \ AND \ \Gamma_k)$ only is meaningful when $\{\Pi_k\}$ and $\{\Gamma_k\}$ are compatible. In my theory, this can be given a meaning in terms of successive measurements; see (5.11).

For a continuous e-variable, like the position ξ of a particle, discussed in Sect. 5.5, a more general construction of the projectors and the corresponding vector spaces must be applied. As mentioned there, the spectral measure E for the operator A for the position ξ is given as follows: For an interval I, $E(I)$ is given by multiplying with the indicator function for the interval. In general $P(\xi \in B) = \text{trace}(\rho E(B))$.

5.7.4   Measurements

We consider again a single observer with an e-variable θ and a corresponding operator $A = \sum_k u_k\Pi_k$. Let the current state be given by a density matrix ρ, and assume non-ideal measurements with data z. A statistical model for these measurements are given by the likelihood $p(z|\theta = u_k)$. Again we have positive operator valued measures, POVM, now defined by

$$M(\{z\}) = \sum_k p(z|\theta = u_k)\Pi_k,$$

$$M(C) = \sum_{z\in C} M(\{z\}).$$

And again the probability distribution of the observations are given by $P(C) = \text{trace}(\rho M(C))$. A corresponding formula is valid also in the continuous case. How this probability model should be analysed, will depend on the concrete situation. I sketch one approach in the next subsection, and give one approach in Sect. 5.13.

5.7.5   Quantum Statistical Inference

In some applications the density matrix ρ depends upon an unknown parameter η. Then the probability measure P above also depends upon η, and we obtain a statistical model. This is the point of departure of Barndorff-Nielsen et al. (2003), where many notions of ordinary statistical inference theory are generalized. It is also the point of departure of Chap. 6 in Holevo (1982).

The probability model for the experiment will be given by

$$P^{\eta}(C) = \text{trace}(\rho^{\eta} M(C)), \tag{5.14}$$

where M is a POVM.

In the spirit of this book it is important to look upon the sets C as sets in the data-space. So (5.14) describes an ordinary statistical model, and all statistical theory applies.

In particular, we can consider symmetries in the data space and corresponding symmetries in the e-variable space; see Sect. 2.2 and also Helland (2004). If a group $\bar{G}$ is defined on the data space, the corresponding group G on the e-variable space is determined by

$$P^{g\eta}(z \in C) = P^{\eta}(z \in \bar{g}C) \ \ g \in G, \ \bar{g} \in \bar{G}. \tag{5.15}$$

The two groups are homomorphic.

In the present setting one can talk about covariant measurements. In analogy with Holevo (1982), Chap. 4, consider the following: Let $\bar{g} \to U(\bar{g})$ be a projective unitary representation of $\bar{G}$ in a Hilbert space H; see the end of Appendix B. The measurement M is called *covariant* if

$$U(\bar{g})^{\dagger} M(C) U(\bar{g}) = M(C_{\bar{g}^{-1}}), \ \ \bar{g} \in \bar{G}, \tag{5.16}$$

for any C, where $C_g = \{z : z = \bar{g}z', z \in C\}$ is the image of the set C under the transformation $\bar{g}$. This gives rise to an interesting theory.

There is a large literature on quantum statistical inference. The field started with the monographs of Helstrom (1976) and Holevo (1982), the latter continued in Holevo (2001). There is much more material in Barndorff-Nielsen et al. (2003) and also in Holevo (1982, 2001). Hayashi (2005) is a collection of papers on the asymptotic theory of quantum statistical inference. A number of recent papers on quantum statistics can be found in Gill et al. (2014).

5.8 Entanglement, EPR and the Bell Theorem

The total spin components in different directions for a system of two spin 1/2 particles satisfy the assumptions of a maximal symmetric epistemic setting. Assume that we have such a system where $j = 0$, that is, the state is such that the total spin is zero. By ordinary quantum mechanical calculations, this state can be explicitly written as

$$|0\rangle = \frac{1}{\sqrt{2}}(|1, +\rangle \otimes |2, -\rangle - |1, -\rangle \otimes |2, +\rangle), \tag{5.17}$$

where $|1, +\rangle \otimes |2, -\rangle$ is a state where particle 1 has a spin component $+1/2$ and particle 2 has a spin component $-1/2$ along the z-axis, and *vice versa* for $|1, -\rangle \otimes |2, +\rangle$. This is what is called an entangled state, that is, a state which is not a direct product of the component state vectors. I will follow my own programme, however, and stick to the e-variable description.

Assume further that the two particles separate, the spin component of particle 1 is measured in some direction by an observer Alice, and the spin component of particle 2 is measured by an observer Bob. Before the experiment, the two observers agree both either to measure spin in some fixed direction a or in another fixed direction b, orthogonal to a, both measurements assumed for simplicity to be perfect. As a final assumption, let the positions of the two observers at the time of measurement be spacelike, that is, the distance between them is so large that no signal can reach from one to the other at this time, taking into account that signals cannot go faster that the speed of light by the theory of relativity.

This is Bohm's version of the situation behind the argument against the completeness of quantum mechanics as posed by Einstein et al. (1935) and countered by Bohr (1935a,b). This discussion is still vigorous in the quantum literature today, although most physicists now support Bohr on the question of completeness of quantum mechanics.

I will be very brief on this discussion here. Let θ be the spin component as measured by Alice, and let η be the spin component as measured by Bob. Alice has a free choice between measuring in the directions a and in the direction b. In both cases, her probability is $1/2$ for each of $\theta = \pm 1$. If she measures $\theta^a = +1$, say, she will predict $\eta^a = -1$ for the corresponding component measured by Bob. According to Einstein et al. (1935) there should then be an element of reality corresponding to this prediction, but if we adapt the strict interpretation of Sect. 4.1 here, there is no way in which Alice can predict Bob's actual real measurement at this point of time. Bob on his side has also a free choice of measurement direction a or b, and in both cases he has the probability $1/2$ for each of $\eta = \pm 1$. The variables θ and η are conceptual, the first one connected to Alice and the second one connected to Bob. As long as the two are not able to communicate, there is no sense in which we can make statements like $\eta = -\theta$ meaningful.

The situation changes. however, if Alice and Bob meet at some time after the measurement. If Alice then says 'I chose to make a measurement in the direction a and got the result u' and Bob happens to say 'I also chose to make a measurement in the direction a, and then I got the result v', then these two statements must be consistent: $v = -u$. This seems to be a necessary requirement for the consistency of the theory. There is a subtle distinction here. The clue is that the choices of measurement direction both for Alice and for Bob are free and independent. The directions are either equal or different. If they should happen to be different, there is no consistency requirement after the measurement, due to the assumed orthogonality of a and b. Note again that we have an epistemic interpretation of quantum mechanics here. At the time of measurement, nothing exists except the e-variables as theoretical quantities and then the observations made by the two observers.

Let us then look at the more complicated situation where a and b are not necessarily orthogonal, where Alice tosses a coin and measures in the direction a if head and b if tail, while Bob tosses an independent coin and measures in some direction c if head and in another direction d if tail. Then there is an algebraic inequality

$$\theta^a \eta^c + \theta^b \eta^c + \theta^b \eta^d - \theta^a \eta^d \le 2. \tag{5.18}$$

Since all the conceptual variables take values ± 1, this inequality follows from

$$(\theta^a + \theta^b)\eta^c + (\theta^b - \theta^a)\eta^d = \pm 2 \le 2.$$

Now replace the e-variables here with actual measurements. Taking then formal expectations from (5.18), assumes that the products here have meaning as random variables; in the physical literature this is stated as an assumption of realism and locality. This leads formally to

$$E(\widehat{\theta^a}\widehat{\eta^c}) + E(\widehat{\theta^b}\widehat{\eta^c}) + E(\widehat{\theta^b}\widehat{\eta^d}) - E(\widehat{\theta^a}\widehat{\eta^d}) \le 2 \tag{5.19}$$

This is one of Bell's inequalities, called the CHSH inequality.

On the other hand, using quantum-mechanical calculations, that is Born's formula, from the basic state (5.17), shows that a, b, c and d can be chosen such that Bell's inequality (5.19) is broken. This is also confirmed by numerous experiments with electrons and photons.

From the point of view of pure epistemic processes the transition from (5.18) to (5.19) is not valid. Without assuming that the estimators $\widehat{\theta}$ and $\widehat{\eta}$ can be defined on some common probability space, one can not take the expectation term by term in Eq. (5.18). The θ's and η's are conceptual variables belonging to different observers. From an epistemic point of view, any valid statistical expectation must take one of these observers as a point of departure. Look at (5.18) from Alice's point of view, for instance. She starts by tossing a coin. The outcome of this toss leads to some e-variable θ being measured in one of the directions a or b. This measurement is an epistemic process, and any prediction based upon this measurement is a new epistemic process. In any inference she should condition upon the choice a or b, since these choices lead to different statistical experiments. This conditioning may be motivated by an extension of the generalized principle of conditioning (GPC), situation (1), where u is the outcome of the coin toss (see Sect. 3.2.3).

By doing predictions from this result, she can use Born's formula. Suppose that she measures θ^a and finds $\theta^a = +1$, for instance. Then she can predict the value of θ^c and hence predict η^c as $-\theta^c$. Thus she can (given the outcome a of the coin toss) compute the expectation of the first term in (5.18). Similarly, she can compute the expectation of the last term in (5.18). But there is no way in which she simultaneously can predict θ^b and η^d. Hence the expectation of the second term (and also, similarly the third term) in (5.18) is for her meaningless. A similar conclusion is reached if the outcome of the coin toss gives b. And of course a similar conclusion is valid if we take Bob's point of view. Therefore the transition from (5.18) to (5.19) is not valid, not by non-locality, but by taking a pure epistemic point of departure. This can also in some sense be called lack of realism: Assuming that it is not meaningful to take expectation from the point of view of an objective observer, then by necessity one must see the situation from the point of view of one of the observers Alice or Bob.

There is a huge literature on Bell's inequality. The first experiments showing a violation of the inequality in quantum mechanical situations were performed by Aspect, but these experiments were criticised, and various authors proposed different 'loopholes' for such experiments. In 2015 three different groups reported 'loophole-free' experiments, and in all these experiments Bell's inequality were violated. For references and further discussion, see Khrennikov (2016b).

Entanglement is very important in modern applications of quantum mechanics, not least in quantum information theory, including quantum computation. It is also an important ingredient in the theory of decoherence; see Schlosshauer (2007), which explains why ordinary quantum effects are not usually visible on a larger scale. Decoherence theory shows the importance of the entanglement of each system with its environment. In particular, it leads in effect to the conclusion that all observers share common observations after decoherence between the system and its environment, and this can then be identified with the 'objective' aspects of the world; which is also what the superior actor D of Sect. 5.1 would find in this situation.

5.9 Mermin's Experiment

Mermin (1985) discusses the following hypothetical experiment to illustrate the peculiar features of quantum mechanics:

Two detectors, one belonging to Alice and one belonging to Bob, are far from each other, and no communication between the two detectors is permitted (spacelike separation). Each detector has a switch that can be set in one of three positions 1, 2 or 3, and each detectors responds to an event by either flashing a green light (G) or flashing a red light (R). Midway between the two detectors there is a source emitting particles, causing simultaneous events at the two detectors.

Alice chooses her switch randomly and so does Bob with his switch. An independent observer reads off the positions of the switches and the responses R or G after each event. For instance, 32RG means that Alice has position 3, Bob position 2, there is a red flash at Alice's detector and a green flash at Bob's detector. A large number of events of this type is read off. After this, the observer notes the following:

1. If one examines only those runs in which both the switches have the same setting, then one finds that the lights always flash the same colors.

2. If one examines all runs, without any regard to how the switches are set, then one finds that the pattern of flashing is completely random. In particular, half the time the lights flash the same color, and half the time different colors.

Is this really possible? asks Mermin, and answers that by classical thinking it is not. Imagine that the detectors are triggered by particles that have a common origin at the source. Suppose, for example, that what each particle encounters as it enters one detector is a target divided into eight regions, labeled RRR, RRG, RGR, RGG, GRR, GRG, GGR, and GGG. Suppose that each detector is wired so that if a particle lands in the GRG bin, the detector flips into a mode in which the light flashes G if the switch is set to 1, R if it is set to 2, and G if it is set to 3; RGG leads to a mode with R for 1 and G for 2 and 3, and so on. One can imagine variants of this, but all such variants leads to an instruction set of this type. The feature 1) will then result for all possible switch settings of the two detectors if and only if both Alice's detector and Bob's detector receives the same instruction.

Can this then be made consistent with the observation 2)? The answer is no. For the purpose of the present argument one can let the probability of each of RRR, RRG,... be arbitrary. Given that the result is RRG, then the detectors will flash the same color when the switches are set to 11, 22, 33, 12, or 21; they will flash different colors for 13, 31, 23. or 32. Thus with this result, the detectors will flash the same color 5/9 of the time. With exactly the same reasoning, for all the results RRG, GRR, RGR, RGG, GRG, and GGR, the detectors will flash the same color 5/9 of the time, since this argument only depends upon the fact that one color appears twice and the other once. But in the remaining cases RRR and GGG, the detectors always flash the same color. Thus by classical thinking, the two detectors will by necessity flash the same color at least 5/9 of the time. This is inconsistent with 2).

The argument just given, corresponds to Bell's inequality for this experiment. The point is now that according to quantum mechanics, Bell's inequality is violated: One can indeed make a quantum mechanical experiment in which both 1) and 2) holds!

Let the source produce two particles of spin 1/2 in the singlet state, that is, with the total spin equal to 0. Let Alice's detector be as follows: If the switch has position 1, she asks for the spin component in the z-direction; if the switch has position 2 or 3, she asks for the spin component in two different directions in a plane orthogonal to the line towards the source, each separated 120° from the z-axis. The detector flashes green if the answer is $+1/2$, red if the answer is $-1/2$. Bob's detector is similar, except that position 1 corresponds to a question in the $-z$-direction, and the directions of his positions 2 and 3 are opposite to Alice's directions corresponding to positions 2 and 3. With this arrangement, it is obvious that 1) will hold always. A straightforward calculation using Proposition 5.7 shows that 2) also holds.

Thus one must by necessity conclude that the classical argument does not hold in the quantum-mechanical setting. What is wrong? According to my view, one must take into account that there are different observers here, and the classical argument must be replaced by an argument from the point of view of one of the observers. First, let us take Alice's point of view. As in the previous section, a valid argument must be conditional on the position chosen by her switch. Also, everything should be conditional on the context. Our task is to find some context where the quantum-mechanical result can be explained.

We describe the context in terms of a third actor Charles. The actions of Charles will be described in very concrete terms. He is assumed to have a box containing four balls, three yellow and one blue. Before each event he draws a ball randomly from the box. This gives the context for the results of Alice and Bob. The context is also constructed in such a way that Alice and Bob always get the same result if their switch is the same. If Alice and Bob have different switches, the context gives them the same result if the ball chosen by Charles is blue, opposite if the ball is yellow. The whole procedure is repeated for every event.

How can this be arranged explicitly? Charles must be assumed to be able to observe the switch choices of Alice and Bob, and must be able to send a message to the two, not about the choice of switches, but about the colour of the ball he has drawn. If Charles observes that the switches are the same, he always reports a blue ball. If he observes that the switches are different, he reports the colour of the ball chosen earlier. Charles does not observe the flash colour at Alice's and Bob's stations.

So let us first look at the experiment from Alice's point of view. To be concrete, assume that she chooses switch 1 and gets as her result a green flash. She does not know Bob's switch position, but she knows the context, and is able to ask for the colour of the ball.

Assume that the answer is 'blue'. After that she knows that if then Bob chooses switch 1, his result will always be a green flash. If he chooses switch 2 or 3, his probability of green will be 1/4 and the probability of red 3/4. If the switch is not recorded, the probability of green will be $1/3 \cdot 1 + 2/3 \cdot 1/4 = 1/2$. Thus the experiment will satisfy 2). She knows from the context that it satisfies 1).

Assume that the answer is 'yellow'. Then she knows that Bob's switch position is 2 or 3, and that the probability of a green ash at Bob's station is 1/2, assuming that at the outset both Alice and Bob have equal probabilities for both colours. Thus the experiment again satisfies 2).

The situation is similar from Bob's point of view.

We need not worry how nature chooses the actor Charles. It is only necessary that one such context produces the result of the quantum experiment. In my view, Mermin's hypothetical experiment clarifies the role of the Bell type inequalities, and the reason why such inequalities can be violated in quantum mechanics.

To the above discussion one may object that no inequality is involved. Zeilinger (2010) discusses a variant of Mermin's experiment leading directly to an inequality under the assumption of local realism.

5.10 The Free Will Theorem

Throughout the times, several authors have proposed various types of hidden variable theories which they claim to be consistent with quantum mechanics. Again and again the scopes of these theories have been limited by so-called no-go-theorems. One of the first and most well-known of these theorems was that of Kochen and Specker (1967): If a theory should be compatible with quantum mechanics, one can not find an arbitrary set of hidden variables that are non-contextual and take definite values at any time.

One of the newest no-go-theorem is The Free Will Theorem of Conway and Kochen (2006, 2008). They take as a point of departure the EPR-type experiment with spin 1 particles, but presumably this can be generalized. They state two assumptions that are weaker than, but implied by quantum mechanics and one assumption which is implied by relativity theory. Under these assumptions they prove:

The Free Will Theorem *If the choice of directions in which to perform spin 1 experiments is not a function of the information accessible to the experimenters, then the responses of the particles are equally not functions of the information accessible to them.*

Thus the particles in a sense have a free will: Their responses are not in any way determined by past history. Past history is here a very wide concept. It can include stochastic variables given in advance, so this kind of simple randomness will not help.

The specific assumptions that Conway and Kochen give for their free will theorem are:

(1) SPIN

Measurements of Alice and Bob are both given in some frame (x, y, z), and the measurements are always 1, 0,1 in some order. This is in particular satisfied by the squares of the spin 1 components along the coordinate axes according to quantum mechanics, which are commuting operators.

(2) TWIN

If the measurements performed by Alice and Bob are along the same axis, they give the same result. This is analogous to what we assumed in the Bell experiment discussed above, only that the signs of Bob's measurements are reversed.

(3) FIN

There is a finite upper bound to the speed with which information can be effectively transmitted. This assumption is weakened in Conway and Kochen (2008).

Admittedly, especially the SPIN-assumption describes a rather special situation, but one can assume that the theorem can be generalized to other situations with entanglement, in the language of the present book: To other situations where Alice and Bob choose their measurement freely, but in different contexts. The result of the Free Will Theorem is then that Nature also chooses its response freely: It is not in any way a function of the past history of the universe. My conjecture is that there is a free will type theorem whenever the physical system observed is complex enough.

What is then meant by a free choice? I adhere to the definition in a short note by Colbeck and Renner (2013): A choice A is free if A is uncorrelated with the set of all events W such that A does not cause W. In the Bell experiment, the choice A made by Alice should be uncorrelated with the past together with anything related to Bob and his observations.

Bob is here any observer which cannot communicate with Alice. If two observers communicate, their free choice is restricted. Then each of them form part of the context experienced by the other person.

5.11   The Schrödinger Equation

During a time when no measurement is done on the system, the ket vector is known in quantum mechanics to develop according to the Schrödinger equation:

$$i\hbar \frac{d}{dt}|\psi\rangle_t = \mathcal{H}|\psi\rangle_t, \qquad (5.20)$$

where $\mathcal{H}$ is a selfadjoint operator called the Hamiltonian (the total energy operator).

I will give two sets of arguments for the Schrödinger equation, one rough and general, and then one specific related to position. The last argument also includes a discussion of the wave function.

5.11.1   The General Argument: Unitary Transformations and Entanglement

Assume that the system at time 0 has a state given by the ket $|\psi\rangle_0$ and at time t by the ket $|\psi\rangle_t$. Let us assume that the contexts are given as follows: We can ask an epistemic question about the variable θ, and the ket corresponding to a specific value of this variable is $|\theta\rangle_0$ at time 0 and $|\theta\rangle_t$ at time t. We have the choice between making an ideal measurement at time 0 or at time t. Since there is no disturbance through measurement of the system between these two time points, the probability distribution of the answer must be the same whatever choice is made. Hence according to Born's formula

$$|_0\langle\theta|\psi\rangle_0|^2 = |_t\langle\theta|\psi\rangle_t|^2. \qquad (5.21)$$

Now we refer to a general theorem by Wigner (1959), proved in detail by Bargmann (1964): If an equation like (5.21) holds, then there must be a unitary or antiunitary transformation from $|\psi\rangle_0$ to $|\psi\rangle_t$. (Antiunitary U means $U^{-1} = -U^\dagger$.)

Since by continuity an antiunitary transformation can be excluded here, so we have

$$|\psi\rangle_t = U_t |\psi\rangle_0$$

for some unitary operator U_t. Writing $U_t = \exp(\frac{A_t}{i\hbar})$ for some selfadjoint operator A_t, and assuming that A_t is linear in t: $A_t = \mathcal{H}t$, this is equivalent to (5.20). In fact, assuming that $\{U_t\}$ forms a strongly continuous group of unitary transformations, the form $\mathcal{H}t$ of A_t follows from a theorem by Stone; see Holevo (2001).

Unitary transformations of states play an important role in quantum mechanics. Both in the continuous case and in the discrete case this can be used to extend the state concept as used in the present book. Note that, subject to linearity, a unitary operator U always can be written as $U = e^{i\mathcal{H}t}$ for some suitable Hamiltonian $\mathcal{H}$, so these transformations can be seen as closely related to time developments of states.

Consider the discrete case. Let the initial state be $|a; k\rangle \otimes |b; j\rangle$, corresponding to the answers of two focused questions: $\theta^a = u_k^a$ and $\gamma^b = v_j^b$. By a unitary transformation, essentially by a time development, this initial state is transformed into a state which can not be written as a product of states in this way, but is a linear combination of such states. This is called an entangled state. In my terminology, entangled states can also be given concrete interpretations: Some fixed time ago they where given as answers to two focused questions. By the inverse unitary transformation, the entangled state may be transformed back to the state $|a; k\rangle \otimes |b; j\rangle$ again. Thus we have a concrete interpretation of the entangled state: Subject to a suitable Hamiltonian, the state can be interpreted as the answer to two focused questions posed at some past time.

5.11.2 Position as an Inaccessible Stochastic Process

As in Sect. 5.3 consider the motion of a non-relativistic one-dimensional particle, but now make time explicit. Since momentum and hence velocity cannot be determined simultaneously with arbitrary accuracy, it is also impossible to determine positions $\xi(s)$ and $\xi(t)$ simultaneously for two different time points s and t with arbitrary accuracy. Hence the vector $(\xi(s), \xi(t))$ is inaccessible. Fix a time point t. Different observers may focus on different aspects from the past of the time t in order to try to predict $\xi(t)$ as well as possible. These aspects may be formulated by propositional logic in different ways, but for reasons discussed in Appendix E I will in this book concentrate on a probabilistic description. Thus observer A_i may predict $\xi(t)$ by conditioning on some σ-algebra $\mathcal{P}_i$ of information from the past. This may be information from some specific time point s_i with $s_i < t$, but it can also take other forms. We must think of these different observers as hypothetical; only one of them can be realized. Nevertheless one can imagine that all possible information, subject to the choice of observer later, is collected in an inaccessible σ-algebra $\mathcal{P}_t$, the past of $\xi(t)$. The distribution of $\xi(t)$, given the past $\mathcal{P}_t$, for each t, can then be represented as a stochastic process.

In the simplest case one can then imagine $\{\xi(s); s \geq 0\}$ as an inaccessible Markov process: The future is independent of the past, given the present. Under suitable regularity conditions, a continuous Markov process will be a diffusion process, i.e., a solution of a stochastic differential equation of the type

$$d\xi(t) = b(\xi(t), t)dt + \sigma(\xi(t), t)dw(t). \tag{5.22}$$

Here $b(\cdot, \cdot)$ and $\sigma(\cdot, \cdot)$ are continuous functions, also assumed differentiable, and $\{w(t); t \geq 0\}$ is a Wiener process. The Wiener process is a stochastic process with continuous paths, independent increments $w(t) - w(s)$, $w(0) = 0$ and $E((w(t) - w(s))^2) = t - s$. Many properties of the Wiener process have been studied, including the fact that its paths are nowhere differentiable. The stochastic differential equation (5.22) must therefore be defined in a particular way; for an introduction to Itô calculus or Stochastic calculus; see for instance Klebaner (1998). One well known result is Itô's formula: For a two times continuously differentiable function f one has:

$$df(\xi(t), t) = f_t(\xi(t), t)dt + f_x(\xi(t), t)d\xi(t) + \frac{1}{2}f_{xx}(\xi(t), t)\sigma^2(\xi(t), t)dt.$$

$$\tag{5.23}$$

Here and in the following, subscripts denote partial derivatives.

There is also the Fokker-Planck equation for the probability density $\rho(x, t)$ of $\xi(t)$:

$$\rho_t(x, t) = -(b(x, t)\rho(x, t))_x + \frac{1}{2}(\sigma^2(x, t)\rho(x, t))_{xx}.$$

So far we have considered observers making predictions of the present value $\xi(t)$, given the past $\mathcal{P}_t$. There is another type of epistemic processes which can be described as follows: Imagine an actor A_j which considers some future event for the particle, lying in a σ-algebra $\mathcal{F}_j$. He asks himself in which position he should place the particle at time t as well as possible in order to have this event fulfilled. In other words, he can adjust $\xi(t)$ for this purpose. Again one can collect the σ-algebras for the different potential actors in one big inaccessible σ-algebra $\mathcal{F}_t$, the future after t. The conditioning of the present, given the future, defines $\{\xi(t); t \geq 0\}$ as a new inaccessible stochastic process, with now t running backwards in time. In the simplest case this is a Markov process, and can be described by a stochastic differential equation

$$d\xi(t) = b_*(\xi(t), t)dt + \sigma_*(\xi(t), t)dw_*(t), \tag{5.24}$$

where again $w_*(t)$ is a Wiener process.

Since t is now running backwards in time, Itô's formula now reads:

$$df(\xi(t), t) = f_t(\xi(t), t)dt + f_x(\xi(t), t)d\xi(t) - \frac{1}{2}f_{xx}(\xi(t), t)\sigma_*^2(\xi(t), t)dt.$$

$$(5.25)$$

The Fokker-Planck equation is now:

$$\rho_t(x, t) = -(b_*(x, t)\rho(x, t))_x - \frac{1}{2}(\sigma_*^2(x, t)\rho(x, t))_{xx}.$$

5.11.3 Nelson's Stochastic Mechanics

Without having much previous knowledge about modern stochastic analysis and without knowing anything about epistemic processes, Nelson (1967) formulated his stochastic mechanics, which serves our purpose perfectly. Nelson considered the multidimensional case, but for simplicity, I will here only discuss a one-dimensional particle. Everything can be generalized. The discussion here will be brief. I refer to Nelson (1967) for details.

Nelson discussed what corresponds to the stochastic differential equations (5.22) and (5.24) with σ and σ_* constant in space and time. Since heavy particles fluctuate less than light particles, he assumed that these quantities vary inversely with mass m, that is, $\sigma^2 = \sigma_*^2 = \hbar/m$. The constant $\hbar$ has dimension action, and turns out to be equal to Planck's constant divided by 2π. This assumes that $\sigma^2 = \sigma_*^2$, a fact that Nelson actually proved in addition to proving that

$$b_* = b - \sigma^2(\ln\rho)_x.$$

Now define

$$u = \frac{1}{2}(b - b_*), \quad v = \frac{1}{2}(b + b_*).$$

Then

$$u = \frac{1}{2}\sigma^2(\ln\rho)_x,$$

and the two Fokker-Planck equations give the continuity equation

$$\rho_t = -(v\rho)_x.$$

By a simple manipulation from this, one finds that

$$u_t = -\frac{1}{2}\sigma^2 v_{xx} - (vu)_x.$$

$$(5.26)$$

Related to (5.23) with (5.22) inserted and (5.25) with (5.24) inserted, Nelson defined the forward and backward derivatives

$$Df(x(t), t) = f_t(x(t), t) + b(x(t), t) f_x(x(t), t) + \frac{1}{2}\sigma^2 f_{xx}(x(t), t);$$

$$D_* f(x(t), t) = f_t(x(t), t) + b_*(x(t), t) f_x(x(t), t) - \frac{1}{2}\sigma^2 f_{xx}(x(t), t),$$

and argued that the acceleration of the particle can be defined by

$$a(t) = \frac{1}{2} D_* D x(t) + \frac{1}{2} D D_* x(t).$$

Then a simple manipulation shows that $D\xi(t) = b(\xi(t), t)$, $D_*\xi(t) = b_*(\xi(t), t)$ and that

$$v_t = a + u u_x - v v_x + \frac{1}{2}\sigma^2 u_{xx}. \tag{5.27}$$

By Newton's law, the force F upon the particle is ma. Assuming that F is derived as the negative gradient of a potential V, we get $a = -m^{-1} V_x$. Inserting this and at the same time $\sigma^2 = \hbar/m$ into (5.26) and (5.27), we have a coupled non-linear set of differential equations for $u(x, t)$ and $v(x, t)$. This can be solved as an initial value problem assuming $u(x, 0) = u_0(x)$ and $v(x, 0) = v_0(x)$ for some given functions u_0 and v_0.

From the relationship between b and b_* we already know that

$$R_x = \frac{m}{\hbar} u,$$

where $R(x, t) = \frac{1}{2}\ln\rho(x, t)$. Let S be defined up to an additive constant by

$$S_x = \frac{m}{\hbar} v,$$

and define the complex function $f(x, t)$ by

$$f = e^{R+iS}.$$

Then $|f(x, t)|^2 = \rho(x, t)$. Nelson interpreted f as the wave function of the particle.

A remarkable fact, noted by Nelson, is that the nonlinear set of Eqs. (5.26) and (5.27) for u and v transforms into a linear equation

$$f_t = i\frac{\hbar}{2m} f_{xx} - i\frac{1}{\hbar} V f + i\alpha(t) f. \tag{5.28}$$

To prove this, we compute the derivatives in (5.28) and divide by f, finding

$$R_t + i S_t = i \frac{\hbar}{2m}(R_{xx} + i S_{xx} + [R_x + i S_x]^2) - i \frac{1}{\hbar} V + i\alpha(t).$$

Taking x-derivatives here and separating real and imaginary parts, we see that this is equivalent to the pair of equations

$$u_t = -\frac{\hbar}{2m} v_{xx} - (vu)_x,$$

$$v_t = \frac{\hbar}{2m} u_{xx} + \frac{1}{2}(u^2)_x - \frac{1}{2}(v^2)_x - \frac{1}{m} V_x.$$

This is the same as (5.26) and (5.27).

Finally, Nelson notes that since the integral of ρ is 1, hence independent of t, if (5.28) holds at all then $\alpha(t)$ must be real. By choosing, for each t, the arbitrary constant in S appropriately, we can arrange for $\alpha(t)$ to be 0. Thus (5.28) is equivalent to

$$i\hbar \frac{\partial}{\partial t} f(x, t) = [\frac{1}{2m}(-i\hbar \frac{\partial}{\partial x})^2 + V(x)] f(x, t). \tag{5.29}$$

This is the Schrödinger equation (5.20) with the Hamiltonian corresponding to the sum of kinetic and potential energy. Note that as in Sect. 5.3 the operator for the momentum of the particle is $-i\hbar \frac{\partial}{\partial x}$.

5.12   The Epistemic Approach to Quantum Theory

In recent years several conferences on quantum foundation have been arranged. It is remarkable that the participants of these conferences disagree strongly on several simple and fundamental questions concerning interpretation. At least two polls have been arranged showing this (Schlosshauer et al. (2013), and Norsen and Nelson (2013)), while a third conference ended up by formulating a list of fundamental open questions (Briggs et al. 2013). In my view, part of the reason for this state of affair is that quantum theory as it exists today is a very formalistic theory, so it is uncertain how it should be interpreted. By contrast, the present book gives a very simple and intuitive point of departure: A quantum state may be determined by the question 'What is the value of θ?' for an e-variable θ, together with the answer to that question. As seen from the discussion of the Bell inequality, the observer plays a role in determining this state, but nevertheless an objective world is created when all real and imagined observers agree. We will see how this simplifies interpretation in three examples usually associated in different ways with the phrase 'quantum paradoxes'.

Example 5.18 (Schrödinger's cat) This is an imaginary situation where a cat is locked inside a sealed box together with some radioactive substance with decay probability 1/2 during the span of time of the experiment, and some poisonous gas is released from a flask which is broken when the radioactive particle decays. The discussion of this example concerns the state of the cat just before the box is opened. Is it half dead and half alive?

To an observer outside the box the answer is simply: He doesn't know. Any accessible e-variable connected to this observer does not contain any information about the death status of the cat. But on the other hand—an imagined observer inside the box, wearing a gas mask, will of course know the answer. The interpretation of quantum mechanics is epistemic, not ontic, and it is connected to the observer. Both observers agree on the death status of the cat once the box is opened.

Example 5.19 (Wigner's friend) The Wigner's friend thought experiment posits a friend of Wigner who performs the Schrödinger's cat experiment after Wigner leaves the laboratory. Only when he returns does Wigner learn the result of the experiment from his friend, that is, whether the cat is alive or dead. The question is raised: Was the state of the system a superposition of "dead cat/sad friend" and "live cat/happy friend," only determined when Wigner learned the result of the experiment, or was it determined at some previous point?

My answer to this is that at each point of time there is one quantum state connected to Wigner's friend and one quantum state connected to Wigner as an observer, depending on the information they have at that time. The superposition given by formal quantum mechanics corresponds to a 'don't know' epistemic state. The states of the two observers agree once Wigner learns the result of the experiment.

Example 5.20 (The two-slit experiment) This is an experiment where all real and imagined observers communicate at each point of time, so there is always an objective state. Particles, either photons or electrons are sent through a screen with two parallel slits towards a second screen. An interference pattern shows up on the second screen even if the particles are sent so slowly that only one particle arrives at a time. The interference pattern disappears when any observer observes which slit each particle goes through.

Look first at the situation where we do not know which slit the particles go through. This is really a 'don't know' situation. Any statement to the effect that the particles somehow pass through both slits is meaningless. The interference pattern can be explained by the fact that the particles are (nearly) in an eigenstate of the component of momentum in the direction perpendicular to the slits in the plane of the slits, namely corresponding to velocity 0 in this direction. If any observer finds which slit each particle goes through, the state changes into an eigenstate for the position component in this direction. In either case the state is an epistemic state for each of the communicating observers, thus also an objective state.

Many physicists find it disturbing that the observer should play an important role in the interpretation of a physical theory. One of the recent conferences on quantum foundation was entitles 'Quantum mechanics without an observer'. However, in my opinion this position is very difficult to hold without arriving at paradoxes. A case of point is that the Bell inequalities have caused very much discussion among physicists. As shown in Sect. 5.8 the apparent paradox around these inequalities can be resolved by taking a purely epistemic point of departure.

So far the epistemic viewpoint has held a minority position among quantum physicists. However, in recent years a new interpretation has arrived at the scene, called Quantum Baysianism or QBism; see Fuchs (2010), von Baeyer (2013) and Fuchs et al. (2013). To cite from Fuchs et al. (2013): 'A QBist takes quantum mechanics to be a personal mode of thought—a very powerful tool that any agent can use to organize his own experience. That each of us can use such a tool to organize our own experience with spectacular success is an extremely important objective fact about the world that we live in. But quantum mechanics does not deal directly with the objective world; it deals with the experiences of that objective world that belongs to whatever particular agent which is making use of the quantum theory.'

Exactly the same words could have been used to describe the interpretation of quantum mechanics advocated in this book, except that I have used the word 'observer' instead of 'agent'. As shown above and shown in the QBism literature, such an interpretation removes the paradoxes and pseudo-problems that have plagued quantum mechanics since its introduction. The difference between the two approaches is that QBism is based upon subjective probabilities; I use basic ideas from statistical inference theory. However, note from Sect. 5.5 that my perfectly rational observer D acts as a Bayesian.

Fuchs (2010) discusses other popular interpretations of formal quantum theory and calls them 'quick fixes'. 'They look to be interpretive strategies hardly compelled by the particular details of the quantum formalism, giving more or less arbitrary appendages to it.'

He uses Robert Spekkens's toy model to motivate the epistemic point of view. As shown in Sect. 4.3, this toy model is closely related to a special case of the maximal symmetrical epistemic setting of Sect. 4.2. Fuchs states that more than two dozen quantum phenomena are reproduced qualitatively by this toy model, and these phenomena can then also be reproduced by my symmetrical epistemic setting. This is confirmed when in Sect. 4.4 the Hilbert space formalism is derived from this setting.

Fuchs regards Bayesianism as a branch of probability theory. As shown in Sect. 2.3, it is really a branch of statistical inference theory. Bayesian theory and practice has become very popular among statisticians in recent years. Many statisticians call themselves Bayesians and claim that the Bayesian formalism should be used in every application of statistics. In this book I take a more pragmatic attitude: Sometimes Bayesian ideas should be used, and sometimes other, so-called frequentist ideas should be used. Both approaches are permitted. This pragmatic attitude is also carried over to the foundation of quantum mechanics. For instance, the probabilities π_k^a in the definition of the density operator in Sect. 5.1 can be Bayesian prior or posterior probabilities, but they can also be derived from frequentist confidence distributions. And the experimental evidence in Sect. 5.4 can in principle be Bayesian or non-Bayesian.

Finally, Fuchs (2010) uses the so-called 'symmetric informationally complete positive operator valued measure' or SIC to integrate and derive the Born formula. I prefer the derivation given in Sect. 5.4 based upon the likelihood principle and the assumption of rationality. It is interesting, however, that both approaches rely on a Dutch book argument. For more about SIC, see the next section.

In all, there are differences between my views and the Quantum Bayesian views. What we share, is a basic epistemic attitude. One place where we differ, is that I allow all three interpretations of probability, the frequentist, the subjective and the symmetry interpretation (see Sect. 2.1.1), while the Quantum Bayesian only concentrates on the subjective interpretation. I also rely in an essential way upon several results related to statistical inference theory.

Another recent treatise advocating an epistemic point of view—in his language a link between realism and information—is Zeilinger (2010). In that book one can also find nice popular discussions of phenomena like quantum computing, quantum teleportation and quantum coding, all closely related to the concept of entanglement.

5.13 More on Quantum Measurements

Recall that the density matrix ρ can be defined by a question 'What is the value of θ?' together with probabilities π_k associated with the possible answers u_k as $\rho = \sum \pi_k |k\rangle\langle k|$. In Sect. 5.7 I addressed the measurement situation where we have data z, and where the question variable was specified. Here we will look at the situation where ρ is completely unknown; in particular, no specified question variable.

First note that if the dimension of the Hilbert space is m, then ρ depends upon $m^2 - 1$ unknown real parameters. This is proved as follows: ρ is in general an Hermitian matrix of trace 1. This implies m real parameters on the diagonal and $m(m - 1)/2$ complex parameters above the diagonal which also determine the parameters below the diagonal, giving a total of $m + 2 \cdot m(m-1)/2 = m^2$ parameter. From this, 1 has to be subtracted because of the trace 1 condition.

Inference on ρ is done by asking new questions 'What is θ^b?' for new known values of b. For simplicity we assume that the e-variables θ^b are maximal, so that the eigenvalues u_j of the corresponding operators A^b are distinct. We now depart from the assumption that the experiment of measuring θ^b is perfect. We assume real data z^b, also taking the values u_j $j = 1, \ldots, m$. The error model $P_{ij} = P(z^b = u_j | \theta^b = u_i)$ is assumed known from earlier experiments with the same measurement apparatus.

Recall that there can be defined a positive operator values measure by the point measures

$$M^b(u_j) = \sum_{i=1}^{m} P_{ij} |b; i\rangle\langle b; i|, \tag{5.30}$$

and that this leads to the following point probabilities in the state defined by ρ:

$$P(z^b = u_j|\rho) = \text{trace}(\rho M^b(u_j)). \tag{5.31}$$

Equation (5.31) gives a model which is linear in the elements of ρ, and linear combinations of these elements can be estimated as follows:

Assume n independent copies of the system described by the density ρ, and perform the measurement b on each of these systems. Assume that y_j of these measurements give $z^b = u_j$ for $j = 1, \ldots, m$. Then $(y_1, \ldots, y_m)$ is multinomial (see Sect. 2.1.2) with parameters n and $\eta_j = P(z^b = u_j|\rho)$. Only $m-1$ of the latter parameters are independent, since the probabilities sum to 1. These probabilities can be estimated by

$$\widehat{\eta}_j = \frac{y_j}{n}, \quad j = 1, \ldots, m-1. \tag{5.32}$$

The covariance matrix of these estimators is given by the diagonal terms $\eta_j(1 - \eta_j)/n$ and the off-diagonal terms $-\eta_i\eta_j/n$. These elements are unknown, but for large n they can be estimated consistently by replacing η_j by $\widehat{\eta}_j$ here.

It is interesting that a central limit theorem is valid for the multinomial distribution. For large n, we have that $\sqrt{n}(\widehat{\eta}_j - \eta_j)$ for $j = 1, \ldots, m-1$ has an approximate multinormal distribution with a covariance given by the diagonal terms $\eta_j(1 - \eta_j)$ and the off-diagonal terms $-\eta_i\eta_j$. Thus for large n we can write $\widehat{\eta}_j = \eta_j + e_j$, where the joint covariance matrix of the small error terms e_j is approximately known. And the e_j's are approximately Gaussian.

By this procedure we are able to estimate $m-1$ linear combinations of the $m^2 - 1$ unknown elements of ρ. To find estimates of all the elements, we must make repeated measurements with different values of b, at least $(m^2-1)/(m-1) = m+1$ values. Suppose that $N = np$ independent copies of the system are available, where $p \geq m+1$, and make p independent experiments with values b_r $r = 1, \ldots, p$. From (5.31) this gives estimates of $\eta_j^r = \text{trace}(\rho M^{b_r}(u_j))$ for $j = 1, \ldots, m-1$ and $r = 1, \ldots, p$.

Since $M^{b_r}(u_j)$ from (5.30) are known operators, this can be written as follows: Let ρ be the vector of the real parameters in ρ written in some order. Let η be the vector of parameters η_j^r written in the natural order. Then we can find a known matrix B such that

$$\eta = B\rho.$$

The dimension of B are $(m-1)p \times (m^2-1)$. It has therefore at least as many rows as columns. Since it contains the arbitrary constants P_{ij}, and since the values of b_r are different (different sets of orthogonal vectors $\{|b; j\rangle\}$), it seems to be a relative weak assumption to assume that is has rank $m^2 - 1$.

From this, it follows that

$$\widehat{\eta} = B\rho + e,$$

where the error vector e is approximately multinormal with an approximately known covariance matrix $\mathbf{\Sigma}$. If $\mathbf{\Sigma}$ is assumed known, statistical theory for linear models leads to the generalized least squares estimate, which also is the maximum likelihood estimate:

$$\widehat{\rho} = (\mathbf{B}^T \mathbf{\Sigma}^{-1} \mathbf{B})^{-1} \mathbf{B}^T \mathbf{\Sigma}^{-1} \widehat{\eta}.$$

This can be translated back to an estimate of ρ. This $\widehat{\rho}$ will be Hermitian of trace 1, but a problem is that it can happen that it is not positive, that is, $\langle \psi | \widehat{\rho} | \psi \rangle$ may be negative for some choice of the ket vector $| \psi \rangle$. A similar problem may occur in statistics, in the estimation of variance components. The problem becomes less if n is large.

A problem which has not been discussed up to now is the choice of the b_r's. Look at the minimal choice of p, $p = m + 1$. Intuitively, all the ket vectors $|b_r; j\rangle$ $j = 1, \ldots, m - 1; r = 1, \ldots, m + 1$ should be as orthogonal as possible. It is of course impossible to have $m^2 - 1$ orthogonal vectors in an m-dimensional space, but one could perhaps relax on the assumption that the $|b_r; j\rangle$ for fixed r and different j should be exactly orthogonal.

A related problem has been discussed in the quantum mechanical literature recently. A SIC POVM (symmetric informationally complete positive-operator-valued measure) is a set of m^2 operators of the form $E_i = (1/m)|i\rangle\langle i|$, where the ket vectors $|i\rangle$ satisfy

$$|\langle i | j \rangle|^2 = \frac{1}{m + 1}.$$

A SIC POVM has been constructed explicitly for a few valued of m. For many more m, there is good numerical evidence that a SIC POVM exists. It has been conjectured that it exists for all values of m, but no proof of this is available up to now; see Wootters (2004) and references there. It is remarkable that Wootters tries to connect the problem of determining an unknown density matrix and the SIC POVM problem to a well known problem from experimental design, that of finding the maximal number of orthogonal Latin squares of a given dimension.

One purpose of the discussion above has been to demonstrate that it may be fruitful to use ideas from mathematical statistics to solve problems in quantum mechanics.

5.14 Discussion

So far, this book falls naturally into two parts, Chaps. 2 and 3 on conventional inference, and the last chapters on quantum mechanics, although the two parts are tied together. Let us first discuss some of the results of the first part.

As indicated in Sect. 3.4, statistical inference is often made in steps. At each step, the results of the previous steps then form a part of the context. And it may initiate more steps. A typical case is when a least squares estimation is done in multiple regression, and this is followed by a residual analysis. Such a sequence is not consistent with the ordinary likelihood principle, but it is consistent with our extended basis.

It is often stated that a weakness with the definitions of sufficiency and ancillarity is that they are strongly model dependent. This can remedied by a stepwise analysis, where new models are tested in steps, following a residual analysis from older models.

Taking more steps in the total inference, our basis is even consistent with algorithmic procedures like Breiman's trees (see Breiman 2001, and references there) and the ordinary partial least squares algorithm see Martens and Næs (1989).

All this must be taken under one proviso, however: The overall goal of the statistical analysis must be formulated first, and taken as part of the context for all the steps.

I have not gone much into formal logic in this book. In Sect. 3.1 I indicated an equivalence between propositional logic and the ordinary basis for probability models. When I now discuss inference in steps, the propositional logic must be extended to temporal logic, for which there is a large literature; see an introduction in Venema (2001). A further extension would be to proceed to first order and higher order logic, but this is beyond the scope of the present book.

As a transition between the two parts of this book, the following remark was made: Statistical literature has much discussion about the way to do inference, but very little on the choice of what to do inference about. These different questions may be conflicting, even complementary. The maximal symmetrical epistemic setting is a way to formalize a situation where only one out of many possible questions may be addressed.

Here are some problems that can be discussed further from the point of view of the present approach to quantum theory.

- The technical Assumption 4.3 of Sect. 4.4.2 is too strong to cover the spin/ angular momentum case. How can it be weakened? The very fact that Theorem 4.1 is valid for the spin/ angular momentum case (see Corollary 5.1 of Sect. 5.2) shows that it indeed must be possible to weaken this condition.
- A closely, related question: Exactly in which cases is it true for a maximal symmetrical epistemic setting that all unit vectors in H are associated with indicators $I(\lambda^a = u_k)$? For spin or angular momentum higher than 1/2, the answer is negative if we limit ourselves to e-variables that are spin- (angular momentum-)components, as can be seen from Corollary 5.2 of Sect. 5.2. On the other hand, the statement holds under all conditions for spin 1/2 particles, as shown in Sect. 5.1.1 and also in Proposition 5.4 of Sect. 5.2. This shows that the statement holds in Hilbert space dimension 2, but that the simple question-and-answer vectors form a smaller set when the dimension is higher. In general, quantum states that are not of this form will emerge by taking linear

combinations, or they can be found a solutions of the Schrödinger equation. Note that every ket vector $|k\rangle$ is trivially the eigenvector of many different operators, so that if at least one of these has a physically meaningful interpretation, one can always associate $|k\rangle$ with at least one question-and-answer pair.

The one-to-one correspondence between indicatiors $I(\theta^a = u_k)$ and certain ket vectors $|a; k\rangle$ under some conditions (Theorem 4.1 etc.) is an important message of this book. As shown in (5.3), this also implies important instances of the superposition principle.

- Can the discussion of Sect. 4.4 be extended beyond the discrete case?
- Can one find examples where one is sure that the rational epistemic setting is indicated in the macroscopic world?
- What about a discussion of open systems? These are thoroughly discussed from a quantum statistical point of view by Holevo (2001).
- Can this approach to quantum mechanics be reconciled with the influential and very deep theory of Richard Feynman; see Brown (2005), or even with its generalization to quantum electrodynamics, see Feynman (1985), and can it be generalized to quantum field theory?
- Can the group-theoretical approach used here in some way throw more light upon elementary particle theory, where group theory is used extensively?
- Can a further development of the discussion here, extended to continuous systems, lead to a reconciliation of quantum theory and relativity theory? It is well known that quantum mechanics can be extended to take into account the special theory of relativity, but that there are conceptual difficulties involved in finding a synthesis between conventional quantum theory and the general theory of relativity. Of course I do not have any solution to these difficulties at present. Already here, however, it is tempting to suggest that gravitational fields and related physical quantities are conceptual variables, and that they are inaccessible inside black holes.

Note that all these questions are connected to my ambitious programme of building a new *foundation* of quantum mechanics on a theory of epistemic processes. The simple idea of an epistemic *interpretation* of quantum mechanics is not affected.

In this book I have tried to give a common foundation of statistics and quantum theory. An important question is then: What is the main feature which make quantum theory essentially different from classical statistical theory? In my opinion this main feature is *complementarity*, a concept that was discussed in detail already by Niels Bohr, and which will be further elaborated in Chap. 6 below. In the setting of the previous chapters, this concept may be approached as follows. Assume that there in some given situation are two different e-variables θ^a and θ^b. Assume further that the vector (θ^a, θ^b) is inaccessible, that is, there is no experiment in which both θ^a and θ^b can be measured accurately. Then θ^a and θ^b are complementary. Examples are spin components in different direction of a single particle; or position and momentum of one particle. Examples abound also in macroscopic situations, like counterfactual questions concerning one person. Note, however, that these macroscopic counterfactual situations cannot easily be linked to quantum theory. Hence there must also be more than complementarity that lies behind quantum probabilities. Exactly what, is at present an open question.

Some clues concerning what is needed in addition to complementarity, can be found in the recent paper by Jaeger (2018). In particular, the notion of *randomness* as defined precisely there, seems to be valuable.

A completely different attempt to find a unified approach to statistics, quantum theory and relativity theory is given by Frieden (1998, 2004). This attempt is much more formal, however, and in my view it does not go enough in depth into the questions discussed.

To emphasize again part of the motivation behind the present book, I cite Hardy and Spekkens (2010): 'Quantum theory is a peculiar creature. It was born as a theory of atomic physics early in the twentieth century, but over time its scope has broadened, to the point where it now underpins all of modern physics with the exception of gravity. It has been verified to extreme high accuracy and has never been contradicted experimentally. Yet despite of its enormous success, there is still no consensus among physicists about what this theory is saying about the nature of reality.'

What I have tried to do here, is to suggest a new approach to the foundation, and thereby a new interpretation, by bringing together basic ideas from statistics and from physics. In the words of John A. Wheeler: 'Science owes more to the clash of ideas than to the steady accumulation of facts.'

The underlying concept of both these sciences is that of an epistemic process: The process of obtaining knowledge about nature from observations. We begin with an epistemic question: 'What is the value of θ?', where θ is some conceptual variable. Then at the end of the process we have some knowledge of θ, in the simplest case complete knowledge; $\theta = u_k$.

In my opinion the approach of the present book may serve to take some of the mysteries off the ordinary formal introduction to quantum theory. A challenge for the future will be to develop the corresponding relativistic theory by using representations of the Poincaré group (the theory of these representations go back to Wigner 1939) and the link to elementary particle physics using the relevant Lie group theory. Group theory is an important ingredient in many parts of physics, and it should be no surprise that it also is relevant for the foundation of quantum mechanics.

My hope is that the present book will contribute to settling the discussions around the interpretation of quantum mechanics. However, it seems probable that this discussion will continue for some time. Arguments for an ontological/ realistic interpretation of the state vector have been put forward by Penrose (2016), Section 2.12. On the other hand, Östborn (2016, 2017) has argued in a detailed way for a quantum model from various epistemic assumptions. Further—in fact very strong—arguments for an epistemic interpretation can be found in the recent development of cognitive models and decision theory in a quantum theory context.

To understand the recent development of quantum cognitive models, one can take the famous example by Tversky and Kahneman (1983) as a point of departure.

A number of people were given a description of a woman, Linda, as having been an outspoken and radical student. Then they were given several options to guess her state now, including feminist (A), bank teller (B), and also feminist and bank teller $(A \wedge B)$. In a number of versions of this experiments, results were given which seem to indicate $P(A \wedge B) > P(B)$, a conclusion which is impossible by using ordinary Kolmogorov probability. This and many similar experiments were explained by Pothos and Busemeyer (2013) by using quantum probability. Op. cit. is a review and discussion paper where a long range of further references may be found. One should also mention the research by Aerts and Gabora (2005a,b) who developed a theory of concepts and their combinations leading to a Hilbert space representation.

Yukalov and Sornette (2014) start by reviewing classical decision theory. Here different prospects are being characterized by their expected utility, the maximization of which leads to the most useful prospect. Thereafter a probabilistic approach is described, in which different actors choose different prospects. Finally, Quantum Decision Theory is proposed, a theory in which conscious operations and subconscious activity work in parallel to determine the prospect probabilities.

This Quantum Decision Theory is developed in detail by Yukalov and Sornette (2010). It is based on the mathematical theory of separable Hilbert spaces. The theory describes entangled decision-making, non-commutativity of subsequent decisions, and intention interference of composite prospects. They demonstrate that all known anomalies and paradoxes, documented in the context of classical decision theory, are reducible to a few mathematical archetypes, all of which find a straightforward explanation in Quantum Decision Theory. A further description of this will lead beyond any theory of epistemic processes, however.

5.15 A General e-Variable Based Approach to Quantum Theory

Consider the world as seen from the point of view of an observer m. This observer focuses upon a question a to ask nature. Such a question can always be formulated in terms of an e-variable θ^a. In much of this book, θ^a has been a discrete variable. However, it could be any variable, real, vector or belonging to any measurable topological space. The general question is: 'What is the value of θ^a?' Let the answer to this question be of the form '$\theta^a \in B$'. The essence of quantum theory, motivated from different points of view in this book, is:

The focused question together with its answer is under certain condition in a one-to-one way associated with a closed subspace V_B^{ma} of some Hilbert space H^m, equivalently with the projector Π_B^{ma} upon this subspace.

When θ^a is of a general form, this theory based upon a finite set of projectors may be too simple. However, the following is always true when the underlying Hilbert space H^m is separable: θ^a *for the subject* m *is under certain technical conditions associated in a one-to-one way with a self-adjoint operator* $A = A^{ma}$ *acting on* H^m. Let the domain of definition of A^{ma} be $D(A^{ma})$. By the spectral theorem for this operator (for a formulation and proof, see for instance Busch et al. 2016), it is always connected to a resolution of the identity, a generalization of the PDI of Sect. 5.7.3, or more precisely, a spectral measure $E = E^{ma}$. Then the question: 'What is the value of θ^a?' together with the answer $\theta^a \in B$ is associated with the projector $E^{ma}(B)$, which is defined on $D(A^{ma})$.

The whole theory is only non-trivial when different questions a and b are non-compatible, that is:

Assume that $a \neq b$ can be found such that the vector of e-variables $\phi = (\theta^a, \theta^b)$ is inaccessible, that is, there is no way in which the question 'What is the value of ϕ?' can be made meaningful for and can be answered completely by the subject m at some given time.

In the words of Niels Bohr, the e-variables θ^a and θ^b are then complementary.

To define a density matrix $\rho = \rho^{ma}$, assume that there is a general measure $d\theta^a$ on the θ^a-space, and that we have an a priori (or otherwise) probability distribution $\pi(\theta^a)d\theta^a$ on this e-variable. Then define $\rho^{ma} = \int \pi(\theta^a)E^{ma}(d\theta^a)$.

Suppose that the observer m wants to measure a new e-variable θ^b, given that his state is defined by the density matrix ρ^{ma}. Then according to Born's formula in a perfect measurement one has $P^{ma}(\theta^b \in B) = \text{trace}(\rho^{ma}E^{mb}(B))$.

In real measurements, one has data z and a probability distribution $p(dz|\theta^b)$. This results in a projective operation valued measure (POVM) $M^{mb}(C) = \int p(C|\theta^b)E^{mb}(d\theta^b)$, and a probability model given the state as $P^{mab}(z \in C) = \text{trace}(\rho^{ma}M^{mb}(C))$. This is from my point of view the starting point of quantum inference theory.

Note that, in the absence of complementarity, this theory reduces to ordinary statistical inference theory (Chap. 2). The e-variable then is a parameter, a fact already discussed in Chap. 1.

Of course there are more aspects of a full epistemic quantum theory. One is the time development given by the Schrödinger equation. Another aspect is the information exchange between different observers m. In Yukalov et al. (2017) the latter questions are discussed from a quantum decision point of view.

References

Aerts, D., & Gabora, L. (2005a). A theory of concepts and their properties I. The structure of sets of contexts and properties. *Kybernetes, 34,* 167–191.

Aerts, D., & Gabora, L. (2005b). A theory of concepts and their properties II. A Hilbert space representation. *Kybernetes, 34,* 192–221.

Aerts, D., de Blanchi, M. S., & Sozzi, S. (2016). The extended Bloch representation of entanglement and measurement in quantum mechanics. *International Journal of Theoretical Physics.* https://doi.org/10.1007/s10773-016-3257-7.

Ballentine, L. E. (1998). *Quantum mechanics: A modern development.* Singapore: World Scientific.

Bargmann, V. (1964). Note on Wigner's Theorem on symmetry operations. *Journal of Mathematical Physics, 5,* 862–868.

Barndorff-Nielsen, O. E., Gill, R. D., & Jupp, P. E. (2003). On quantum statistical inference. *Journal of the Royal Statistical Society B, 65*, 775–816.

Barut, A. S., & Raczka, R. (1985). *Theory of group representation and applications.* Warsaw: Polish Scientific Publishers.

Bing-Ren, L. (1992). *Introduction to operator algebras.* Singapore: World Scientific.

Bohr, N. (1935a). Quantum mechanics and physical reality. *Nature, 136*, 65.

Bohr, N. (1935b). Can quantum-mechanical description of physical reality be considered complete? *Physical Review, 48*, 696–702.

Breiman, L. (2001). Statistical modeling: The two cultures. *Statistical Science, 16*, 199–231.

Briggs, G. A. D., Butterfield, J. N., & Zeilinger, A. (2013). The Oxford Questions on the foundation of quantum physics. *Proceedings of the Royal Society A, 469*, 20130299.

Brown, L. M. (Ed.) (2005). *Feynman's thesis: A new approach to quantum theory.* New Jersey: World Scientific.

Busch, P. (2003). Quantum states and generalized observables: A simple proof of Gleason's Theorem. *Physical Review Letters, 91*(12), 120403.

Busch, P., Lahti, P. J., & Mittelstaedt, P. (1991). *The quantum theory of measurement.* Berlin: Springer.

Busch, P., Lahti, P., Pellonpää, J.-P., & Ylinen, K. (2016). *Quantum measurement.* Berlin: Springer.

Caves, C. M., Fuchs, C. A., & Schack, R. (2002). Quantum probabilities as Bayesian probabilities. *Physical Review, A65*, 022305.

Colbeck, R., & Renner, R. (2013). A short note on the concept of free choice. arXiv: 1302.4446 [quant-ph].

Conway, J., & Kochen, S. (2006). The free will theorem. *Foundations of Physics, 36*, 1441–1473.

Conway, J., & Kochen, S. (2008). The strong free will theorem. arXiv: 0807.3286 [quant-ph].

Einstein, A., Podolsky, B., & Rosen, N. (1935). Can quantum-mechanical description of physical reality be considered complete? *Physical Review, 47*, 777–780.

Everett, H. III (1973). The theory of the universal wave function. In N. Graham, B. DeWitt (Eds.), *The many worlds interpretation of quantum mechanics.* Princeton: Princeton University Press.

Feynman, R. P. (1985). *QED: The strange theory of light and matter.* Princeton: Princeton University Press.

Frieden, B. R. (1998). *Physics from fisher information: A unification.* Cambridge: Cambridge University Press.

Frieden, B. R. (2004). *Science from fisher information: A unification.* Cambridge: Cambridge University Press.

Fuchs, C. A. (2010). QBism, the Perimeter of Quantum Bayesianism. arXiv: 1003.5209v1 [quant-ph].

Fuchs, C. A., Mermin, N. D., & Schack, R. (2013). An introduction to QBism with an application to the locality of quantum mechanics. arXiv: 1311.5253v1 [quant-ph].

Gill, R., Guta, M., & Nussbaum, M. (2014). New horizons in statistical decision theory. *Mathematisches Forschungsinstitut Oberwolfach. Report No. 41.*

Giulini, D. (2009). Superselection rules. arXiv: 0710.1516v2 [quant-ph].

Griffiths, R. B. (2014). The consistent history approach to quantum mechanics. In E. N. Zalta (Ed.), *Stanford encyclopedia of philosophy.* Stanford: Metaphysics Research Lab, Stanford University.

Griffiths, R. B. (2017a). What quantum measurements measure. *Physical Review A, 96*, 032110.

Griffiths, R. B. (2017b). Quantum information: What is it all about? *Entropy, 19*, 645.

Hammond, P. J. (2011). Laboratory games and quantum behavior. The normal form with a separable state space. Working paper. Dept. of Economics, University of Warwick.

Hardy, L., & Spekkens R. (2010). Why physics needs quantum foundations. arXiv: 1003.5008 [quant-ph].

Hayashi, E. (Ed.) (2005). *Asymptotic theory of quantum statistical inference. Selected papers.* Singapore: World Scientific.

Helland, I. S. (2004). Statistical inference under symmetry. *International Statistical Review, 72*, 409–422.

Helland, I. S. (2006). Extended statistical modeling under symmetry; the link toward quantum mechanics. *Annals of Statistics, 34*, 42–77.

Helland, I. S. (2008). Quantum mechanics from focusing and symmetry. *Foundations of Physics, 38*, 818–842.

Helland, I. S. (2010). *Steps towards a unified basis for scientific models and methods*. Singapore: World Scientific.

Helstrom, C. W. (1976). *Quantum detection and estimation theory*. New York: Academic Press.

Holevo, A. S. (1982). *Probabilistic and statistical aspects of quantum theory*. Amsterdam: North-Holland.

Holevo, A. S. (2001). *Statistical structure of quantum theory*. Berlin: Springer-Verlag.

Jaeger, G. (2018). Developments in quantum probability and the Copenhagen approach. *Entropy, 20*, 420–438.

Khrennikov, A. (2016b). After Bell. arXiv: 1603.086774 [quant-ph].

Klebaner, F. C. (1998). *Introduction to stochastic calculus with applications*. London: Imperial College Press.

Kochen, S., & Specker, E. P. (1967). The problem of hidden variables in quantum mechanics. *Journal of Mathematics and Mechanics, 17*, 59–87.

Lehmann, E. L., & Casella, G. (1998). *Theory of point estimation*. New York: Springer.

Ma, Z.-Q. (2007). *Group theory for physicists*. New Jersey: World Scientific.

Martens, H., & Næs, T. (1989). *Multivariate calibration*. Hoboken, NJ: Wiley.

Mermin, N. D. (1985). Is the moon there when nobody looks? *Physics Today, 38*, 38–47.

Messiah, A. (1969). *Quantum mechanics* (Vol. II). Amsterdam: North-Holland.

Murphy, G. J. (1990). *C*-algebras and operator theory*. Boston: Academic Press.

Nelson, E. (1967). *Dynamical theories of Brownian motion*. Princeton: Princeton University Press.

Norsen, T., & Nelson, S. (2013). Yet another snapshot of fundamental attitudes toward quantum mechanic. arXiv:1306.4646v2 [quant-ph].

Penrose, R. (2016). *Fashion, faith, and fantasy in the new physics of the universe*. Princeton: Princeton University Press.

Peres, A. (1993). *Quantum theory: Concepts and methods*. Dordrecht: Kluwer.

Pothos, E. M., & Busemeyer, J. R. (2013). Can quantum probability provide a new direction for cognitive modeling? With discussion. *Behavioral and Brain Sciences, 36*, 255–327.

Schlosshauer, M. (2007). *Decoherence and the quantum-to-classical transition*. New York: Springer.

Schlosshauer, M., Kofler, J., & Zeilinger, A. (2013). A snapshot of fundamental attitudes toward quantum mechanics. *Studies in History and Philosophy of Modern Physics, 44*, 222–238..

Tversky, A., & Kahneman, D. (1983). Extensional versus intuitive reasoning: The cojunction fallacy in probability judgements. *Psychological Review, 90*, 293–315.

Vedral, V. (2011). Living in a quantum world. *Scientific American, 304*(6), June 2011, 20–25.

Venema, Y. (2001). Temporal logic. In L. Goble (Ed.), *The Blackwell guide to philosophical logic*. Hoboken, NJ: Blackwell.

von Baeyer, H. C. (2013). Quantum weirdness? It's all in your mind. *Scientific American, 308*(6), June 2013, 38–43.

von Neumann, J. (1927). Wahrscheinlichkeitstheoretischer Aufbau der Quantenmechanik. *Nachrichten von der Gesellschaft der Wissenschaften zu Göttingen, Mathematisch-Physikalische Klasse* 1927, 245–272.

Wigner, E. (1939). On unitary representations of the inhomogeneous Lorentz group. *Annals of Mathematics, 40*, 149–204.

Wigner, E. P. (1959). *Group theory and its application to the quantum mechanics of atomic spectra*. New York: Academic Press.

Wootters, W. K. (2004). Quantum measurements and finite geometry. arXiv:quant-ph/0406032v3.

Xie, M., & Singh, K. (2013). Confidence distributions, the frequentist distribution estimator of a parameter - a review. Including discussion. *International Statistical Review, 81*, 1–77.

Yukalov, V. I., & Sornette, D. (2010). Mathematical structure of quantum decision theory. *Advances in Complex Systems, 13*, 659–698.

Yukalov, V. I., & Sornette, D. (2014). How brains make decisions. *Springer Proceedings in Physics, 150*, 37–53.

Yukalov, V. I., Yukalova, E. P., & Sornette, D. (2017). Information processing by networks of quantum decision makers. arXiv: 1712.05734 [physics.soc-ph].

Zeilinger, A. (2010). *Dance of the Photons: From Einstein to quantum teleportation.* New York: Farrar, Straus and Giroux.

Östborn, P. (2016). A strict epistemic approach to physics. arXiv:1601.00680v2 [quant-ph].

Östborn, P. (2017). Quantum mechanics from an epistemic state space. arXiv:1703.08543 [quant-ph].

Chapter 6
Macroscopic Consequences

Abstract Philosophical consequences of the view of science and in particular the view of quantum theory expressed in this book are discussed. In particular the relationship between science and religion is briefly touched upon. Culture is seen as a part of the context for making decisions. A thorough discussion of the concept of complementarity, also extended to macroscopic settings, is given.

6.1 Philosophical Considerations

To repeat the views of this book: The quantum formulation can be seen as having to do not with how nature is, but with our process of obtaining knowledge about nature. We focus on certain questions to nature, and obtain answers to those focused questions. I indicate in different ways that essential parts of the quantum formulation can be derived by considering such a process—an epistemic process. These are relatively deep results which require some mathematics to derive.

Who is this 'we' who ask questions and obtain answers? It can be a single observer or a group of communicating observers. The epistemic process that this (these) observers perform(s) can be likened to statistical inference in some way, and the quantity which he/she (they) ask questions about and obtain information on, can be likened to a statistical parameter. I have introduced a new name, an e-variable, to cover both parameters in statistics and these physical quantities.

I claim that the basic principles of statistics, the conditionality principle, the sufficiency principle and the likelihood principle can be generalized to all inference on parameters and to all simple e-variables connected to experiments.

I go on and derive the Born formula from a version of the likelihood principle together with an assumption of rationality. Also, the Schrödinger equation is derived by an assumption of observers taking into account both σ-algebras in the past and in the future.

I stress that my theory is not a hidden variable theory, although it bears some resemblance with such a theory. The inaccessible variable ϕ is not a hidden variable, but a mathematical variable upon which group actions may be defined. The e-variables are not hidden variables, but closely connected to the epistemic processes. (Note that the parameters of statistics exist only in our minds.) Also:

© Springer-Verlag GmbH Germany, part of Springer Nature 2018

I. S. Helland, *Epistemic Processes*, https://doi.org/10.1007/978-3-319-95068-6_6

Doing inference has to do with intuitive processes in the brain, and the brain is no computer and cannot be simulated by any system of computers, so the epistemic process which lie in the foundation here, cannot be simulated by any system of computers.

Although essentially new arguments behind quantum mechanics are presented in this book, I still regard conventional Hilbert space based quantum mechanics as extremely useful when it comes to calculations. It has developed very far since its beginning in the previous century; see for instance a modern book like Ballentine (1998). It still has a very vigorous development; see the many articles posted each month in arXiv:quant-ph or articles published in many good journals. It is not the purpose of this book to try to change this culture. I only claim that an alternative, and perhaps more intuitive basis can be found.

In the same way, the statistical culture, as it is described in Chap. 2, has a vigorous development today. What I try to point out, is, that these two cultures may be seen to have a common basis in the concept of an epistemic process. Nevertheless, I am quite sure that neither quantum mechanics as a science nor mathematical statistics as a science nor applied statistics as a tool in many empirical sciences would have developed as far as they have if some sort of a synthesis between the two cultures had been taken place from the beginning. This may be linked to the quantum mechanical concept of complementarity. Universality and creativity may in some sense be seen as complementary qualities.

The concept of complementarity is extremely important in this book. Up to now it has mainly been connected to the process of obtaining knowledge, that is, the epistemic process. As just seen, it can also be associated with human abilities. Humans observe, make decisions and act. The complementarity concept can also be connected to the decisions and the actions. Assume for instance that a student is to work on some given assignment. He can have his focus on satisfying teacher A or on satisfying teacher B in the decisions and actions he make when working on the assignment. These can be complementary foci. The complementarity can be reduced if more time and concentration is devoted to the assignment.

There are other, complementary, approaches to quantum mechanics than the one through epistemic processes. Some of these were mentioned in Sect. 4.1. Of particular interest are the approaches by Hardy (2001, 2011, 2012, 2013). In a series of papers, Wetterich (2008a,b, 2009, 2010a,b,c,d) has explored the relationship between classical statistical ensembles and quantum mechanics.

Also for mathematical statistics there are other, complementary, approaches, for instance through ordinary decision theory.

Going back to the epistemic process situation, the basic feature of the approach in Chap. 4 was focusing: Ask a selected focused question to nature and obtain a specific answer. From a given observer's point of view this defines a state of nature. The remaining assumptions on the group actions introduced there are mainly to make the derivation of the ordinary Hilbert space apparatus under certain conditions possible.

The focusing used in Assumption 4.1 was precise and formal. Informally, focusing is very often necessary in our daily life when we want to obtain knowledge before making decisions and acting from these decisions. We simply do not have the capacity to absorb all the knowledge from all the sources that we are confronted with.

An epistemic process as used in this book is a very wide concept. As stated in Sect. 3.4, every epistemic process involves decisions, a decision to ask a question and a decision to accept the answer. It is interesting to note that also the other assumptions made in this book have informal analogues for humans making decisions.

1. The rational epistemic setting and the arguing leading to the Born formula was derived from assuming: (a) The perfectly rational actor D. When making decisions, most humans will have ideals which they look up to, ideals to the effect of being as rational as they are able to be. (b) The focused likelihood principle. When making decisions, most humans will try to use all relevant available data and if possible also use their prior model for the situation. These are the important elements which they rely upon.
2. The Schrödinger equation was discussed in detail only for the case of a one-dimensional position, but it was stated that it can be generalized. The assumptions made in this derivation were related to an inaccessible stochastic processes where the observers were able to condition both on the past events and on the future events for accessible focusing from this process. This has an analogue the situation a human is in before and after his actions when planning what decisions to do next. He then takes into account both past events and possible future events.

Thus if we stretch our imagination a little, the two time developments of the quantum state, which have caused so much discussion in the physical literature, can both be said to be connected to mechanisms related to the decision making processes of an observer.

It may also be interesting to speculate around the fact that the free will theorem follows from the assumptions of quantum mechanics, admittedly in a special case, but it is possible that it is valid for most situations that are complicated enough. Humans are governed by their free will, and they are constantly confronted with other humans that are governed by *their* free will.

So what should one mean about the question of reality? It is obvious that the moon is there when nobody looks. (See the title of Mermin 1985). In general *any physical system has an existence which is independent of all observers, and it exists even when there is no observer at all.* It is the state of the system—whatever that means—which we have limited ability to obtain information of.

All these speculations, and indeed the whole idea of a purely epistemic foundation of quantum mechanics, make one a little uneasy, however. The universe was created 13.8 billion years ago, and physical laws, including quantum mechanics have presumably been valid since then. How can then everything be so tightly connected to the human observer? This is of course an obvious question; I nevertheless thank Bill Wootters, oral communication, for mentioning this point to me.

A possible solution in the spirit of this book, is connected to the imagined observers of Assumption 5.4 (Sect. 5.1) and the perfectly rational actor D of Sect. 5.5. To go into more details on the question on what these elements stand for, is again mere speculation, but at least all assumptions made up to now are consistent with the following world view, which also provides a link to the ontic interpretation of quantum mechanics. A similar proposal was in due time made by the philosopher George Berkeley in order to avoid accusations of solipsism. (Bent Selchau, personal communication.)

For several reasons I have chosen to believe that there is a Creator of the universe, who during the creation also observed it. Then later this Creator is at each time able to observe, make decisions and act. He is perfectly rational.

I believe that we humans are created in His image, but we are imperfect. The last statement is obvious; the first may be argued for from the fact that we also are able to observe, make decisions and act. This is the basis for the idea of an epistemic process.

With now a divine Creator entering the stage, this idea obtains a new dimension. If the idealized Heavenly observers agree on the result of *their* epistemic process, this result must be said to be an objective fact. This gives us a simple tentative argument through the concept of an epistemic process, which was argued above to lie behind the formalism of quantum mechanics, to an ontic view of the world. (And all attempts to find an ontological foundation of quantum mechanics can be seen as attempts to see the world from the perspective of the Heavenly Creator.)

Such an ideal epistemic process can only be imperfectly mimicked by human observers. Nevertheless, when several of us agree on an observation, we can be fairly sure that this is an objective fact. Such a conclusion is strengthened if the epistemic process leading to the conclusion is a scientific investigation.

The divine Creator may be called God. He is worshiped in different ways in different cultures and He is seen in different ways by different humans. This can be explained by the fact that we humans only can have an imperfect image of God. It is also connected to the fact that we all have different contexts, also when making deep decisions. The concept of a context has played an important role in this book, it is important in any epistemic process and it is important for any process of making decisions.

From my perspective, the ultimate actor God must in some way be the same across all cultures, and He must be acting over and above what particular image each single person might have of Him or of aspects of Him.

At the outset we know little about the goals behind the decisions made by God. My own conviction is related on Albert Einstein's saying: The Lord is subtle, but not malicious. It is also based upon the God who ends the play Brand of Henrik Ibsen: Deus caritatis; the God of love. Thus there seems in my opinion to be a God which is good and wants the best for us humans. But in a world where we all have our free will, God is confronted with many complementary goals. On the one hand He is almighty. On the other hand He seems to meet logical impossibilities if He should do the best for absolutely all of us.

By all these speculations I have been entering the realm of theology, which is not my speciality. However, one can never stop wondering about the large and difficult questions. Some of the answers must remain open at this stage.

Thus I value high very many aspects of religion. However, I have great difficulties with the attitude: 'We in our religious community are right. The others are wrong.'

I am strongly against any kind of fundamentalism. Extreme Muslims may become terrorists in the belief that they have the right religion. Israeli settlers occupy Palestinian land in the belief that they have the right religion. It is important to have an open mind towards the beliefs of other people, but such an open mind should also have its limitation. There are no simple solutions to the deep conflicts in this world.

In the same way as I believe that there exists an ultimate God, I also believe that there in some sense is an ultimate science. It is very fruitful to do science in various scientific cultures, but there must be a logical way to understand the conclusions obtained in different cultures in a unified way. This is a personal conviction behind the work of this book, but the view may perhaps be generalized to other human activities.

At this point it is natural to stress that also science has its limitations. Science is not able to explain consciousness. Science is not able to grasp in any way the spiritual power behind a symphony by Ludvig von Beethoven, a painting of Pablo Picasso, the finding of theorems by Nils Henrik Abel or the finding of theories by Richard Feynman.

We all have a mind which can not be scrutinized in detail by any scientific investigation, however far our knowledge of the brain is developed. Thus there is a room for a dimension in life that goes beyond science, in my view, also a room for religion. A further discussion of my views on science and religion can be found in Helland (2017).

6.2 More on the Nature of the Superior Actor

We all go through our lives making repeated decisions in different contexts. These decisions are governed by our free will, but they may also be influenced by people that we look up to, who perhaps have done similar decisions before. In our childhood, the persons that form our basis are most often our parents, but later other ideals may take over. Human beings that suffer from a confused relation to their first ideals, may later have difficulties in making good decisions, and they may end up with having psychological problems. Much mental illnesses can be explained in this way.

As scientists we also have ideals that we look up to. These may be personal, or they may be substantiated through certain well-defined principles. In Sect. 5.5 I made the assumption that the experimentalist A, when posing a focused question to nature, made his decisions inspired by an ideal D, and that D was perfectly rational. This may be regarded as a simplification. In reality, when making our decisions, we are influenced by a multitude of conscious or subconscious sources. All these sources are here collected together in the actor D. I assume that D has a positive influence on A, positive with respect to the goal that A has, in this case the question that A has chosen as the focus of his experiment.

Let us look at the process of making decisions in some greater generality. People in different cultures make their decisions partly intuitively on the basis of cultural values. These values may have a historical origin, and they may also be related to religion. Christianity, Islam and Judaism are all founded upon the belief in a personal God. The believers act under the assumption that there is a God behind everything, and that God is perfect. They believe at the same time that He influences all human beings, also those who serve as ideals for others. In this sense, God may take the role as the ultimate ideal D within the relevant culture.

In general a culture may be looked upon as part of a man's context when making his decisions. At the outset, all human beings should be respected, and so also the context they have for making their choices. Hence it is a part of my philosophy that no culture should in principle be seen as definitely better than other cultures when it comes to inspiring people's decisions. However, this tolerance has it limits; one of these is an ultimate respect for people's life. Extremists taking lives under the belief that their own culture is threatened by other cultures, should of course not in any way be accepted. But in addition there are other universal ethical rules that should be respected.

In essence certain cultural values and more generally certain value-contexts for making decisions may be seen from a global point of view to be more satisfactory than other set of values, but this can only be determined by rational arguments. Hence communication between cultures is very important in our world as it is now. As a particular continuation of this statement, this book in itself is written with the partial purpose of finding a common language with which one can communicate across scientific epistemic cultures.

6.3   Quantum Mechanics, Decisions, and Complementarity

This Section can be read independently of the rest of the book.

The modern technological development would have been extremely difficult without scientific theories. In a certain sense physics lies behind all natural science, and it is impossible to discuss modern physics without touching quantum mechanics in some way or other.

The great American physicist Richard Feynman said once: 'If somebody claims that he understands quantum mechanics, he lies.' This statement is still valid, but during the recent years new elements of understanding have appeared.

The crucial point is that quantum mechanics is a formalism, a set of calculating rules for how one can predict the outcome of experiments. These calculating rules have had an enormous success; they have been used for everything from small elementary particles to complex chemical and biological systems, and in every case the predictions have been 100% in agreement with the results of experiments.

However, the great question is how one shall interpret these calculating rules. Here the physicists disagree, also today. During the recent years there has been held a long range of international conferences on the foundation of quantum mechanics. A great number of interpretations have been proposed; some of them look very peculiar to the laymen. For instance, during one period it was popular to assume that there exist millions or billions of parallel worlds, and that a new world appears every time when one performs a measurement. Some take this point of view even today.

On two of these conferences recently there was taken an opinion poll among the participants. It turned out to be an astonishing disagreement on many fundamental and fairly simple questions. One of these questions was: Is the quantum mechanics a description of the objective world, or is it only a description of how we obtain knowledge about reality? The first of these descriptions is called ontological, the second epistemic. (From Webster's Unabridged Dictionary: epistemic: what concerns or comes from knowledge, or the conditions for obtaining knowledge.) A similar, but not quite identical distinction is realistic versus non-realistic. Up to now most physicists have supported the ontological or realistic interpretation of quantum mechanics, but versions of the epistemic interpretation have received a fresh impetus during the recent years.

One such version is QBism, or quantum-Bayesianism. The predictions of quantum mechanics involve probabilities, and a QBist interpret these as purely subjective probabilities, attached to a concrete agent, or observer. There are many elements of QBism which represents something completely new, both in relation to classical physical theory, in relation to many peoples conceptions of science in general and also in relation to earlier interpretations of quantum mechanics. The essential thing is that the observer plays a role which cannot be eliminated. The comprehension of reality for a person differ from person to person, at least at a given point of time, and this is in principle everything that can be said, at a given point of time.

According to QBism there is no other reality than the subjective one attached to each single agent. This statement must be made precise to be understood correctly. Firstly, one talks about an ideal agent, and secondly, groups of agents which communicate mutually, can go in and act as one agent as long as one talks about one measurements. When all potential ideal agents agree about an observation, this observation is a real property of the world.

Nevertheless, these are aspects of physics—and science—which can be surprising for many people, but in my opinion such viewpoints may be necessary, not only in physics, but also in many other areas of life.

Such an understanding of reality can in my opinion be made valid for very many aspects of reality. We humans can have a tendency to experience reality differently. Partly, this can be explained by the fact that we give different meaning to the concepts we use. Or we can have different contexts for our appreciations. An important aspect is that we focus differently.

Subjective Bayes-probabilities have also been in fashion among groups of statisticians. Personally, I mean that it can be very fruitful to look for analogies between statistical inference theory and quantum mechanics, but then one must look more broadly upon statistics and statistical inference theory, not only focus on subjective Bayesianism. This is only one of several philosophies that can form a basis for statistics as a science. Referring to Sects. 6.1 and 6.2, I will assume that the superior actor D acts as a Bayesian, but for human observers also other philosophies must be allowed.

My point of departure is that I look upon a quantum state as the result of two decisions: A decision to focus upon a question to nature, and a decision to interpret the answer. The present section is also about focusing in human decisions in general, simple decisions, more complicated decisions and even deeper decisions which can concern philosophical questions. We all go through life and make decision after decision, make choice after choice.

It can be of interest to look upon how some ideas from modern physics can illuminate these processes. It must be emphasized that this an account of my own opinions, which are far from shared by all physicists. However, Niels Bohr, nearly hundred years ago, expressed similar thoughts, admittedly not quite as radical as this.

My own view upon quantum mechanics is inspired by the QBism, but I mean that one more generally should take as a point of departure a fundamental theory for epistemic processes, processes with the purpose of obtaining knowledge about something. In my opinion a theory of such processes could play a role both for our understanding of daily life and for our understanding of science, quantum mechanics in particular.

Epistemic processes, at least the simplest of them, involve decisions in two stages: First a decision to choose a focus. Then collection of data, and finally an informed decision about what these data say about the phenomenon that we have focused upon. Traditional decision theory is only concerned with the last one of these decisions.

It is very important to find a good enough theory of human decisions. What dominates science today, is a far developed—but in its basis relatively simple—theory for decisions under uncertainty. This covers much of economic theory and also statistical inference theory, and is thus an important part of the foundation both of economy as a science and of statistics as a science.

But human decisions are very complex. Firstly, the decisions may depend on the order in which we want to do our decisions; more generally most decisions will depend on a context, partly determined by earlier decisions. Secondly, many decisions may be a result of a complicated interaction between the conscious and the subconscious. This is not covered by traditional decision theory.

During the recent 5–6 years there has been proposed and developed a new formal theory for decisions, where these are in part conscious and in part subconscious. This theory is inspired by quantum mechanics (Yukalov and Sornette 2008, 2009, 2010, 2011, 2014). There has also in the recent years appeared aspects of the sciences economy and psychology which are analogous to certain aspects of basic quantum theory. (See Khrennikov 2010.) This can all be coupled to decisions.

Decisions can be made by single persons or by groups of people. A group of people can agree to go collectively into a decision process, and can make collective decisions on which actions should be done after the process is finished. All decisions and all actions—whether done on a single person level or on a group level, should to the best of one's ability be done in an intelligent way, where one takes into account all accessible knowledge. For many decisions this may take time.

But in certain case we do not have so much time for our decisions. This is true for most practical decisions taken in everyday life. An important example taken from daily life is that of driving a car. Here one must take quick decisions and at every point of time focus on other car drivers, pedestrians, bicyclists, traffic signs etc.. To be able to do this, concentration is important, but it is also important to have good training. Other cases where we have to take quick decisions, are in verbal communication with other people.

All decisions are made in a context. This context may be purely physical, but it can also be historically determined or be tied to the personality of the one who takes the decision. If it is a question of a conscious decision, it can be critical to know what concept the person has at his disposal in order to formulate his thoughts.

Our opinions can depend on our background, what we have experienced earlier and what persons we have been communicating with or have been influenced by. But at the same time we have free will to take decisions, in particular to formulate our viewpoints and opinions.

In discussing these and similar questions, it can be useful to look at the quantum mechanical concept *complementarity*. For a thorough discussion of complementarity in physics, see Plotnitsky (2013). The concept was originally introduced by Niels Bohr to describe what it is possible to measure physically, but in various talks Bohr also looked upon extensions of the complementarity concept. Such extensions are also of great current interest.

First look upon the purely physical aspect. It turns out that in principle it is impossible to simultaneously measure the velocity and position of a particle. Velocity and position are complementary quantities. It turns also out that this problem is less for particles with large mass, i.e., heavy particles. Thus the degree of complementarity in this sense is largest for light particles. Concretely this is expressed in Heisenberg's uncertainty relation: The product of the uncertainty in velocity and the uncertainty in position is greater or equal to Planck's constant divided by the mass of the particle. If we try to measure the position accurately, we disturb the system so much that it is impossible to measure the velocity accurately. And similarly if we try to measure the velocity accurately.

The terms complementary and complementarity have several meanings, both in physics and elsewhere. In this section I will let these concepts refer to two or more aspects of reality which are difficult or impossible to grasp or to have an attitude to at the same time, but where both (all) in some way are needed to get a complete picture of reality.

In this way the accurate measurement of velocity and the accurate measurement of position are complementary *activities*, and the corresponding *quantities* are also complementary. Many physicists are skeptical to using these concepts outside the concrete physical context, but we will see that it can be fruitful.

Here is Plotnitsky's definition of complementarity:

(a) a mutual exclusivity of certain phenomena, entities, or conceptions; and yet
(b) the possibility of applying each one of them separately at any given point; and
(c) the necessity of using all of them at different moments for a comprehensive account of the totality of phenomena that we consider.

This definition points at the physical situation discussed above, and has Niels Bohr's interpretation of quantum mechanics as a point of departure. However, in my opinion the definition can also be carried over to a long range of macroscopic phenomena or conceptions.

A simple example: A student works with a difficult assignment. One goal can be to get it finished fast; another goal can be to hand in a paper which is as good as possible. These are clearly complementary goals. If his ability to concentrate is good, this may to some extent reduce the degree of complementarity.

Opinions and viewpoints of different persons may often be complementary. Examples of this can be seen daily in newspaper debates. But the differences in opinions may go deeper, and have their basis in complementary *world views*.

Good authors can write novels where the reader understands each single person's descriptions of reality, even in cases where these descriptions are not fully compatible. In such cases one can of course discuss if there is an objective reality behind these world views. It may be that the author's point is just that it is not very fruitful to look for such a complicated objective reality. In any case, to find such an objective reality, one may have to go beyond the conceptual basis for each single person. This is similar in many political conflicts, where the schism is between groups of people.

In quantum mechanics, the concept of reality has played a prominent role in recent discussions. In QBism, the emphasis is moved to each person's experience of reality. In this section I want to go one step further and talk about the collective experience of each group of communicating persons. For different persons or for different groups of communicating persons, their perceptions of reality may be complementary. This may be true when observing the microworld.

Many physics papers discuss two actors Alice and Bob, being so far away from each other that they do not communicate. All physicists agree that there exist situations where the observations of Alice and Bob are entangled. From an epistemic point of view the two actors may also have complementary comprehensions of the world because they focus differently. According to the physicist John A. Wheeler, each observer can create his/her own history.

I claim that this may be equally true for persons—or groups of persons—making experiences in the macroworld. People may tend to have different—complementary—world views.

The summer 2014, in the middle of the Gaza war, both the Israeli and the Palestinian ambassador to Norway were interviewed in a major Norwegian newspaper about the situation in the Middle East. The two had clearly complementary world views.

The Israeli ambassador talked about safety for his population and about Hamas using human shields for their launching of rockets. He also referred to holocaust and stressed that the Jews had strong reasons for seeking their own land. He also mentioned that Hamas had broken several cease-fire agreements, and emphasized that Hamas would not recognize Israel as a sovereign state.

The Palestinian ambassador described the long term occupation of the West Bank and the brutal attacks on the Gaza stripe. He emphasized strongly all the humiliations that the Palestinians are and have been met with. He mentioned illegal Israeli settlements and Apartheid-like conditions in Israel. Of course he also talked about the many civilian losses during the war, especially losses of children.

We, while hearing all this, can of course form our own opinions. But to what extent are these opinions dependent upon which information we by chance have obtained, and not least, upon which group we belong to? It is a fact that Danes largely are more Israel-friendly than Norwegians. Demark would not sign a common Nordic resolution about the Middle East war, but chose instead a more watered down EU-resolution.

Without doubt, our opinions in general can be influences by which country we happen to live in, by which period of time we live in, and more generally, by which culture we belong to.

Sometimes, in such situations, it can be useful to step back and just say that the two world views are complementary. This can be linked to trying to respect both parties, something which is important if one should happen to be in a position where one can help in peace negotiations.

Nevertheless, to only be neutral and rest on the complementarity concept can be dangerous. As humans we have both the right and the duty to take a definite stand on various questions. This is the great logical dilemma that we are faced with in every situation: To make a conscious decision while we at the same time know that we are guided by unknown subconscious causes. In reality this is a problem in every case where we shall make a difficult decision. However, for many people this dilemma is not a big problem: One makes decisions on an intuitive basis.

The complementarity concept must not make us into value-relativists. There is something right and wrong in this world. During the last world war it was right to dissociate oneself from Hitler and his fellows, and we must definitely dissociate ourselves from the human cleansings of Stalin and Pol Pot. In general we should dissociate ourselves from everybody who do not respect fundamental human rights, and there are also other moral issues where we can say something of absolute validity. My own conviction is that moral questions of this kind could be approached through a faith in God, but unfortunately there are many problems where such a conviction do not lead to simple, unique solutions.

To take a concrete and very actual problem: What should we mean about the European refugee issue? From an ideal point of view our borders should be more open, when seeing all the suffering among the refugees. But at the same time there is a limitation on how many refugees the various countries have a capacity to receive in a decent way. This is an area where the public debate has been hard recently. To a certain extent the debate has been dominated by complementary points of view. This should not prevent us from making up our own opinions.

Politics is not simple, and it should not be simple. Many people in Western countries were unambiguously enthusiastic when the Arabic spring started, overthrowing regime after regime. But unfortunately, the spring has turned into winter in many of the affected countries.

Both in daily life, in politics and in science it can be necessary to focus. Another essential point is that all decisions are made in a context. Finally, one should be able to communicate verbally all conscious decisions.

Focusing can be done at all stages of the decision process. As complementary goals, complementary activities and complementary world views are concerned, it can be a solution to focus upon one of the goals, one of the activities and one of the world views. But in particular in the last case, a more intelligent and creative solution can be to try to find a partial synthesis.

Look again on the Middle East conflict, and the two ambassadors who had complementary world views. For many westerners it has been important to take a clear and unambiguous standpoint for one or the other party in this conflict. But as a sensible person has said: This is not a soccer game. There are serious issues at stake for both parties. A more constructive question can be: What can reduce the degree of complementarity? What obstacles get in our way for at the end to reach a lasting two state solution? Or for other lasting solutions? As emphasized in the previous two sections, I do not think that any solution can be found without taking religion and cultural aspects of religion into account.

Concepts from our understanding of modern physics can contribute to enlightening difficult problems, both concerning our own decisions, decisions by groups of people, and world views lying behind serious conflicts. One hope should be that a correct understanding of science may create a conceptual apparatus giving both scientists, politicians and others in leading positions inspiration to work for good human purposes, in its final consequence to work for peace here on earth. A clear view on interpretations of modern physics, and extensions of such interpretations to other areas, can in my opinion contribute to making such a conceptual apparatus. Of course, this does not mean that this contribution from modern physics is a unique contribution for peace. Many good people in our societies work for a similar final goal, and many good people pray for a solution. One point with this section has been to explain how such a work can be partly motivated completely rationally—with a basis in a possible interpretation of the most rational of all sciences: Fundamental physical theory.

References

Ballentine, L. E. (1998). *Quantum mechanics: A modern development.* Singapore: World Scientific.

Hardy, L. (2001). Quantum theory from five reasonable axioms. arXiv: 0101012v4.[quant-ph].

Hardy, L. (2011). Reformulating and reconstructing quantum theory. arXiv: 1104.2066v1 [quant-ph].

Hardy, L. (2012). The operator tensor formulation of quantum theory. arXiv: 1201.4390v1 [quant-ph].

Hardy, L. (2013). Reconstructing quantum theory. arXiv: 1303.1538v1 [quant-ph].

Helland, I. S. (2017). The conception of God as seen from research on the foundation of quantum mechanics. *Dialogo Journal, 4*(1), 259–267.

Khrennikov, A. (2010). *Ubiquitous quantum structure*. Berlin: Springer.

Mermin, N. D. (1985). Is the moon there when nobody looks? *Physics Today, 38*, 38–47.

Plotnitsky, A. (2013). *Niels bohr and complementarity: An introduction*. New York: Springer.

Wetterich, C. (2008a). Probabilistic observables, conditional correlations, and quantum physics. arXiv: 0810.0985v1 [quant-ph].

Wetterich, C. (2008b). Quantum entanglement and interference from classical statistics. arXiv: 0809.2671v1 [quant-ph].

Wetterich, C. (2009). Zwitters: particles between quantum and classical. arXiv: 0911.1261v2 [quant-ph].

Wetterich, C. (2010a). Quantum particles from coarse grained classical particles in phase space. arXiv: 1003.3351v1 [quant-ph].

Wetterich, C.(2010b). Probabilistic time. arXiv: 1002.2593v1 [quant-ph].

Wetterich, C. (2010c). Quantum particles from classical probabilities in phase space. arXiv: 10003.0772v1 [quant-ph].

Wetterich, C. (2010d). Quantum mechanics from classical statistics. *Annals of Physics, 325*, 852–884.

Yukalov, V. I., & Sornette, D. (2008). Quantum decision theory as a quantum theory of measurement. *Physics Letters A, 372*, 6867–6871.

Yukalov, V. I., & Sornette, D. (2009). Processing information in quantum decision theory. *Entropy, 11*, 1073–1120.

Yukalov, V. I., & Sornette, D. (2010). Mathematical structure of quantum decision theory. *Advances in Complex Systems, 13*, 659–698.

Yukalov, V. I., & Sornette, D. (2011). Decision theory with prospect interference and entanglement. *Theory and Decision, 70*, 383–328.

Yukalov, V. I., & Sornette, D. (2014). How brains make decisions. *Springer Proceedings in Physics, 150*, 37–53.

Appendix A
Proof That the Generalized Likelihood Principle Follows from the GWCP and the GWSP (Birnbaum's Theorem); The Discrete Case

Let E_1 and E_2 be the two experiments in the generalized likelihood principle, and let E^* be the mixed experiment from the GWCP. On the sample space of E^* define the statistic

$$t(j, z_j) = \begin{cases} (1, z_1^*) \text{ if } j = 1 \text{ and } z_1 = z_1^* \text{ or if } j = 2 \text{ and } z_2 = z_2^* \\ (j, z_j) \text{ otherwise} \end{cases}.$$

I will use the factorization theorem to prove that $t(j, z_j)$ is a sufficient statistic in the mixed experiment E^*. Define

$$h(j, z_j | \tau) = \begin{cases} c \text{ if } (j, z_j) = (2, z_2^*) \\ 1 \text{ otherwise} \end{cases},$$

where c is the constant of proportionality between the two likelihoods. Define for both values of j and for all z_j:

$$g(t|\theta, \tau) = g((j, z_j)|\theta, \tau) = f^*((j, z_j)|\theta, \tau),$$

where f^* is the point probability in E^*.

Now for all sample points except $(2, z_2^*)$ (but including $(1, z_1^*)$), we have $t(j, z_j) = (j, z_j)$, so

$$g(t(j, z_j)|\theta, \tau)h(j, z_j|\tau) = g((j, z_j)|\theta, \tau) \cdot 1 = f^*((j, z_j)|\theta, \tau).$$

© Springer-Verlag GmbH Germany, part of Springer Nature 2018
I. S. Helland, *Epistemic Processes*, https://doi.org/10.1007/978-3-319-95068-6

Such a factorization also holds for $(2, z_2^*)$. Namely, by using the definitions of t, h, g, f^* together with the GWCP, we have

$$g(t(2, z_2^*)|\theta, \tau)h(2, z_2^*|\tau) = g((1, z_1^*)|\theta, \tau)c = f^*((1, z_1^*)|\theta, \tau)c = c\frac{1}{2}f_1(z_1^*|\theta, \tau)$$

$$= c\frac{1}{2}L_1(\theta|z_1^*, \tau) = \frac{1}{2}L_2(\theta|z_2^*, \tau) = \frac{1}{2}f_2(z_2^*|\theta, \tau) = f^*((2, z_2^*)|\theta, \tau).$$

Here L_1 and L_2 are the likelihoods of the two experiments E_1 and E_2, and I have used the premise of the generalized likelihood principle.

Thus by the factorization theorem, $t(j, z_j)$ is a sufficient statistic for θ, and by the GWSP we have that the evidence about θ in E^* given by $(1, z_1^*)$ and $(2, z_2^*)$ are the same. By the GWCP, $(1, z_1^*)$ gives the same evidence as z_1^* in E_1 and $(2, z_2^*)$ gives the same evidence as z_2^* in E_2. Hence these latter evidences must also be the same, and the generalized likelihood principle follows.

Appendix B
Some Group Theory, Operator Theory and Group Representation Theory

A group G is defined in mathematics as a set of elements g with a composition $g_1 g_2$ satisfying the axioms: (1) There is a unit e such that $eg = ge = g$ for all g; (2) For each g there is an inverse g^{-1} such that $g^{-1} g = g g^{-1} = e$; (3) The composition is associative: $(g_1 g_2) g_3 = g_1 (g_2 g_3)$ for all g_1, g_2, g_3.

The group is Abelian (commutative) if $g_1 g_2 = g_2 g_1$ for all g_1, g_2.

Important examples of groups are the additive group on the real numbers and the multiplicative group on the positive real numbers.

For a finite group G, each element $g \in G$ has an order m such that $g^m = e$. A cyclic subgroup is the group with elements $\{e, g, g^2, \ldots, g^{m-1}\}$. In general a group G may have several subgroups $G' \subseteq G$.

Most of the groups used in this paper are group actions, that is, transformation groups on some set Φ, even though some must be seen as abstract groups. A transformation g of Φ is any function on the elements $\phi \in \Phi$ which is one-to-one and onto. These functions can be composed by $(g_1 g_2)(\phi) = g_1(g_2(\phi))$, and they have inverses g^{-1}. The existence of a unit and the associative law are automatic. Thus by definition they form a group. For any set Φ the group of all transformations on Φ exists, and is called the automorphism group of Φ. Thus many groups G of this paper may be considered as subgroups of some automorphism group.

An orbit of a transformation group G is a subset of Φ, the set of all ϕ that are transformed from a single element ϕ_0, that is $\{\phi : \phi = g \phi_0$ for some $g \in G\}$. The restriction of G to an orbit or to a set of orbits will itself be a group transformation, which again without possible confusion can be called G. Restrictions to orbits of groups on the parameter space were used in connection with model reduction in Sect. 2.2 and later. This constraint on model reduction is important if the same transformation group shall be kept during the reduction. Such reductions were important in connection to the maximal symmetrical epistemic setting used in introducing the quantum mechanical perspective.

A group where the only orbit is the full group, is said to be transitive. For a transitive group, each element of Φ can be transformed to each other element by some group action.

© Springer-Verlag GmbH Germany, part of Springer Nature 2018
I. S. Helland, *Epistemic Processes*, https://doi.org/10.1007/978-3-319-95068-6

The stabilizer of an element $\phi_0 \in \Phi$ is the subgroup H of G such that $h(\phi_0) = \phi_0$ for $h \in H$. If $\phi_1 = g\phi_0$, then $H(\phi_1) = gH(\phi_0)g^{-1}$. For some groups the stabilizer is trivial.

Let in general both the set Φ and the group G be given some topology, both spaces assumed to be locally compact. Then one can under quite general conditions (see Helland (2010) or any mathematical text on this) define in a unique way (except for a multiplicative constant) two positive measures, a left Haar measure μ_G satisfying $\mu_G(gD) = \mu_G(D)$ and a right Haar measure ν_G satisfying $\nu_G(Dg) = \nu_G(D)$ for all $g \in G$ and all Borel sets $D \subseteq G$.

Then turn to invariant measures on the set Φ itself. In mathematical texts, Φ is often itself treated as a group, and then the concepts of Haar measures carry over. But this is not satisfactory for all statistical applications. In Helland (2010, Subsection 3.3 and Appendix A.2.2) a summary of a way to fix this is given. In that book, group actions were written to the right: $\phi \to \phi g$ so that $\phi(g_1 g_2) = (\phi g_1)g_2$. This is uncommon, but has a certain logical advantage: In the product $g_1 g_2$, we get that g_1 is applied first and then g_2.

A left-invariant measure on Φ is then any measure μ satisfying $\mu(g(B)) = \mu(B)$ for any $g \in G$ and for any Borel set $B \subseteq \Phi$, while a right-invariant measure is any measure ν satisfying $\nu(Bg) = \nu(B)$ for all g, B. In Helland (2010, Theorem A1) it was proved that a right invariant measure always exists on a given orbit of G if the stabilizer of one element, hence of all elements, of this orbit is compact. This is the case under weak technical assumptions (Wijsman 1990, proper group actions; see) if G is locally compact. In Helland (2010, Subsection 3.3) a list of arguments were given why the right invariant measure should be used as an objective prior in statistics if such a prior is required.

The invariant measure is unique up to a multiplicative scalar if the group action is transitive, otherwise invariant measures can be introduced independently on each orbit. For compact groups and in many other cases the left-invariant measure and the right-invariant measure can be taken as identical. When Φ is compact, the invariant measure can be taken as normalized: $\nu(\Phi) = 1$.

Two groups G and R are homomorphic if there exists a function T from G to R such that $T(g_1 g_2) = T(g_1)T(g_2)$ for all g_1, g_2 and such that $T(e) = e'$, the unit in R. Then also $T(g^{-1}) = T(g)^{-1}$. They are isomorphic if T is one-to-one. Then they may be considered as the same abstract group.

Next let us introduce some basic algebra. A vector space is an Abelian group under addition where also multiplication by scalars is defined. In this paper we mainly consider finite-dimensional complex vector spaces, meaning that the scalars are complex numbers and that there exists a set of basis vectors e_i; $i = 1, \ldots, k$ that are linearly independent: $\sum_i c_i e_i = 0$ implies $c_1 = \ldots = c_k = 0$. A linear operator A on a vector space is a function from the vector space into itself satisfying

$$A(c_1 \boldsymbol{a}_1 + c_2 \boldsymbol{a}_2) = c_1 A\boldsymbol{a}_1 + c_2 A\boldsymbol{a}_2.$$

By relating it to the basis vectors, a linear operator can always be represented by a square matrix:

$$Ae_j = \sum_i e_i D(A)_{ij}.$$

Then if $a = \sum_j e_j a_j$ and $b = Aa = \sum_i e_i b_i$, we have $b_i = \sum_j D(A)_{ij} a_j$. Thus if $\underline{a}$ is the column vector of components a_j and similarly for $\underline{b}$, we get $\underline{b} = D(A)\underline{a}$; in complete analogy to $b = Aa$.

If a is represented by the column vector $\underline{a}$, we define $a^\dagger$ as represented by the row vector $(a_1^*, \ldots, a_k^*)$, where $*$ denotes complex conjugate. The scalar product $a^\dagger b$ is defined as $\sum_i a_i^* b_i$. This scalar product is linear in the second vector and antilinear in the first, in agreement with the tradition in physics. Mathematicians tend to use a scalar product which is linear in the first vector. This is in effect just a cultural difference, but to outsiders it is annoying.

Two vectors a and b are orthogonal if $a^\dagger b = 0$. With this interpretation, the basis vectors e_i are automatically pairwise orthogonal and have norm $\|e_i\| \equiv \sqrt{e_i^\dagger e_i} = 1$. In general one can always find many sets of n orthogonal basis vectors in an n-dimensional vector space.

A vector space with the structure above is called an inner product space. This notion can be generalized to infinite-dimensional spaces, having an infinite set of basis vectors. The norm $\|a\| = \sqrt{a^\dagger a}$ induces a metric, hence a topology on this space by $d(a, b) = \|a - b\|$. The space is complete in this metric if $\|a_n - a_m\| \to 0$ $(n, m \to \infty)$ implies that there exists an a such that $\|a_n - a\| \to 0$. A complete inner product space is called a Hilbert space. A closed subspace of a Hilbert space is again a Hilbert space. A finite-dimensional inner product space is always complete, hence a Hilbert space.

The identity operator I is defined by $Ia = a$, and the multiplication of operators by $(AB)(a) = A(Ba)$. Then (in the finite-dimensional case) $D(I)$ is diagonal with 1's on the diagonal, and $D(AB) = D(A)D(B)$, ordinary matrix multiplication. An operator A is invertible if there exists an A^{-1} such that $A^{-1}A = AA^{-1} = I$. A finite-dimensional operator A is invertible if and only if $\det(D(A)) \neq 0$; then $D(A^{-1}) = D(A)^{-1}$.

The conjugate of an operator A, $A^\dagger$, is defined by $a^\dagger(Ab) = (A^\dagger a)^\dagger b$. Slightly different, but equivalent notations for scalar products and conjugates, using kets and bras, were used in the main text. An operator A is called Hermitian if $A^\dagger = A$. An operator V is called unitary if $V^{-1} = V^\dagger$.

An eigenvector v and an eigenvalue θ are solutions of $Av = \theta v$. An operator A is Hermitian if and only if all its eigenvalues are real-valued. Eigenvectors corresponding to different eigenvalues are then automatically orthogonal. In the k-dimensional Hermitian case there are always sets of k pairwise orthogonal eigenvectors.

A group representation of G is a continuous homomorphism from G to the group of invertible linear operators V on some vector space H:

$$V(g_1 g_2) = V(g_1) V(g_2).$$

It is also required that $V(e) = I$, the identity. This assures that the inverse exists: $V(g)^{-1} = V(g^{-1})$. The representation is unitary if the operators are unitary $(V(g)^\dagger V(g) = I)$. If the vector space is finite-dimensional, we have a representation $D(V)$ on the square, invertible matrices. For any representation V and any fixed invertible operator K on the vector space, we can define a new representation by $W(g) = K V(g) K^{-1}$. One can prove that two equivalent unitary representations are unitarily equivalent, so K can be chosen as a unitary operator.

A subspace H_1 of H is called invariant with respect to the representation V if $u \in H_1$ implies $V(g)u \in H_1$ for all $g \in G$. The null-space $\{0\}$ and the whole space H are trivially invariant; other invariant subspaces are called proper. A group representation V of a group G in H is called irreducible if it has no proper invariant subspace. A representation is said to be fully reducible if it can be expressed as a direct sum of irreducible subrepresentations. A finite-dimensional unitary representation of any group is fully reducible. In terms of a matrix representation, this means that we can always find a $W(g) = K V(g) K^{-1}$ such that $D(W)$ is of minimal block diagonal form. Each one of these blocks will represent an irreducible representation. They are all one-dimensional if and only if G is Abelian. The blocks may be seen as operators on subspaces of the original vector space, the irreducible subspaces. These are important in studying the structure of the group.

A useful result is Schur's Lemma (see for instance Barut and Raczka 1985):

Let V_1 and V_2 be two irreducible representations of a group G; V_1 on the space H_1 and V_2 on the space H_2. Suppose that there is a transformation T from H_1 to H_2 such that

$$V_2(g)T(v) = T(V_1(g)v)$$

for all $g \in G$ and $v \in H_1$.

Then either T is zero or it is an isomorphism. Furthermore, if $H_1 = H_2$, then $T = \lambda I$ for some complex number λ.

Let ν be the right and left invariant measure of the space Φ induced by the group G, assuming the two to be equal, and consider the Hilbert space $H = L^2(\Phi, \nu)$. Then the right regular representation of G on H is defined by $U^R(g)f(\phi) = f(\phi g)$ and the left regular representation by $U^L(g)f(\phi) = f(g^{-1}\phi)$. These representations always exist, and they can be shown to be unitary.

If V is an arbitrary representation of a compact group G in H, then there exists in H a new scalar product defining a norm equivalent to the initial one, relative to which V is a unitary representation of G.

For references to some of the vast literature on group representation theory, see Helland (2010, Appendix A.2.4).

A generalization of a unitary representation is a projective unitary representation. The unitary matrices then satisfy

$$U(g)U(h) = \omega(g, h)U(gh),$$

where the complex multiplier ω satisfies

$$|\omega(g, h)| = 1,$$

$$\omega(g, h)\omega(gh, k) = \omega(g, hk)\omega(h, k),$$

$$\omega(g, e) = \omega(e, g) = 1,$$

with e being the unit element of the group G. The discussion of Sect. 4.4 can be generalized to such representations. For a transitive group G let $\phi = h\phi_0$, and define the left regular representation by $U(g)f(\phi) = \omega(g, g^{-1}h)f(g^{-1}\phi)$. This will provide an arbitrary phase factor for the state vectors.

Appendix C
Proof of Two Results Related to Quantum Mechanics

Proof of Theorem 5.3 of Sect. 5.3 Let $\epsilon > 0$ be given. Find first $a > 0$ so large that $\int_{-\infty}^{-a} |f(\xi)|^2 d\xi$, $\int_a^{\infty} |f(\xi)|^2 d\xi$, $\int_{-\infty}^{-a} |\xi f(\xi)|^2 d\xi$, $\int_a^{\infty} |\xi f(\xi)|^2 d\xi$ all are less than $\epsilon/4$. Assume that n is so large that $\xi_{n1} < -a$ and $\xi_{nk_n} > a$. Since f is uniformly continuous on $[-a, a]$ it follows that $\int_{-a}^a |f_n(\xi) - f(\xi)|^2 d\xi \to 0$, so $\| f_n - f \| \to 0$. Since $1 - \sum_j I_{nj}$ is less than the indicator of $(-\infty, -a]$ plus the indicator of $[a, \infty)$, we have $\int |\xi f(\xi) - \xi f(\xi) \sum_j I_{nj}(\xi)|^2 d\xi < \epsilon/2$. Now

$$\xi f(\xi) \sum_j I_{nj}(\xi) - A_n f_n(\xi) = \sum_j (\xi f(\xi) - \xi_{nj} f_n(\xi_{nj})) I_{nj}(\xi).$$

Hence using the uniform continuity of $k(\xi) = \xi f(\xi)$ on $[-a, a]$, we get $\int |\xi f(\xi) - A_n f_n(\xi)|^2 \to 0$.

Proof of the Direct Part of Proposition 5.5 of Sect., 5.4 Define $a_k = p(z^a|\tau, \theta^a = u_k^a)$ and $b_k = cp(z^b|\tau, \theta^b = u_k^b)$. Then (5.8) is

$$\sum_k a_k |a; k\rangle\langle a; k| = \sum_k b_k |b; k\rangle\langle b; k|.$$

In these sums we can collect together terms with equal coefficients. Let $P_j = \sum_{k \in C_j} |a; k\rangle\langle a; k|$, where C_j is defined such that $a_k = a'_k$ when $k, k' \in C_j$, similarly define Q_j on the righthand side. Redefine a_j as a_{k_j} whenever $k_j \in C_j$, and redefine b_j similarly. Then

$$\sum_j a_j P_j = \sum_j b_j Q_j, \tag{C.1}$$

$\{P_j\}$ and $\{Q_J\}$ are orthogonal sets of projection operators, $a_j \neq a_{j'}$ whenever $j \neq j'$ and $b_j \neq b_{j'}$ whenever $j \neq j'$. We can order the terms in (C.1) such that $a_1 > a_2 > \ldots > 0$ and $b_1 > b_2 > \ldots > 0$. Furthermore we can multiply each side of

© Springer-Verlag GmbH Germany, part of Springer Nature 2018
I. S. Helland, *Epistemic Processes*, https://doi.org/10.1007/978-3-319-95068-6

(C.1) with itself any number of times, giving

$$\sum_j a_j^m P_j = \sum_j b_j^m Q_j$$

for $m = 1, 2, \ldots$. When m is large enough, the first term on each side of this equation will dominate completely, and we must have $a_1^m P_1 = b_1^m Q_1$ for all large enough m. But since P_1 and Q_1 are projection operators, this is only possible if $a_1 = b_1$ and $P_1 = Q_1$.

Then we can subtract the term $a_1 P_1$ from the lefthand side of (C.1), subtract the equal term $b_1 Q_1$ from the righthand side of the equation, and repeat the argument. It follows that $a_j = b_j$ and $P_j = Q_j$ for each j, which is the conclusion of the proposition.

Appendix D
Proof of Busch's Theorem
for the Finite-Dimensional Case

The main point of the proof is to show that any generalized probability measure
on effects extends to a unique positive linear functional on the vector space of all
bounded linear Hermitian operators. This is done in steps.

1. It is trivial that $\mu(E) = n\mu(\frac{1}{n}E)$ for all positive integers. It follows that
   $\mu(pE) = p\mu(E)$ for all rational numbers in $[0, 1]$. By approximating from
   below and from above by rational numbers, this implies that $\mu(\alpha E) = \alpha\mu(E)$
   for all real numbers α in $[0, 1]$.
2. Let A be any positive bounded operator in H. Then there is a positive number
   α such that $\langle u|Au \rangle \leq \alpha$ for all unit vectors u. Then E defined by $E = (1/\alpha)A$
   is an effect. Thus we can always write $A = \alpha E$ for an effect E. Assume now
   that there are two effects E_1 and E_2 such that $A = \alpha_1 E_1 = \alpha_2 E_2$. Assume
   without loss of generality that $\alpha_2 > \alpha_1 > 0$. Then by (1) $\mu(E_2) = \frac{\alpha_1}{\alpha_2}\mu(E_1)$, so
   $\alpha_1\mu(E_1) = \alpha_2\mu(E_2)$. Therefore we can uniquely define $\mu(A) = \alpha_1\mu(E_1)$.
3. Let A and B be positive bounded operators. Take $\gamma > 1$ such that $\frac{1}{\gamma}(A+B)$ is an
   effect. Then we can write $\mu(A+B)$ as $\gamma\mu(\frac{1}{\gamma}(A+B)) = \gamma\mu(\frac{1}{\gamma}A) + \gamma\mu(\frac{1}{\gamma}B) = \mu(A) + \mu(B)$.
4. Let C be an arbitrary bounded Hermitian operator. Assume that we have two
   different decompositions $C = A - B = A' - B'$ into a difference of positive
   operators. Then $A + B' = A' + B$ implies $\mu(A) + \mu(B') = \mu(A') + \mu(B)$.
   Hence $\mu(A) - \mu(B) = \mu(A') - \mu(B')$, so we can uniquely define $\mu(C)$ as
   $\mu(A) - \mu(B)$. It follows then easily from (3) that $\mu(C + D) = \mu(C) + \mu(D)$
   for bounded Hermitian operators.
5. This is extended directly to $\mu(C_1 + \ldots + C_r) = \mu(C_1 + \ldots + C_{r-1}) + \mu(C_r) = \mu(C_1) + \ldots + \mu(C_r)$ for finite sums.

Let $\{|k\rangle; k = 1, \ldots, n\}$ be a basis for H. Then for any Hermitian operator C
we can write $C = \sum_{i,j} c_{ij}|i\rangle\langle j|$, where c_{ij} are complex numbers satisfying $c_{ij}* = c_{ji}$. Define the operator ρ by $\rho_{ij} = \mu(|i\rangle\langle j|)$. Then ρ is a positive operator since

© Springer-Verlag GmbH Germany, part of Springer Nature 2018
I. S. Helland, *Epistemic Processes*, https://doi.org/10.1007/978-3-319-95068-6

$\langle v | \rho v \rangle = \mu(|v\rangle\langle v|)$ for any vector $|v\rangle$. Also

$$\text{trace}(\rho) = \sum_i \rho_{ii} = \sum_i \mu(|i\rangle\langle i|) = \mu\left(\sum_i |i\rangle\langle i|\right) = \mu(I) = 1,$$

so ρ is a density operator.

We have $\mu(C) = \sum_{i,j} \rho_{ij} c_{ij} = \text{trace}(\rho C)$, and this holds in particular when C is an effect.

Appendix E
Propositional Logic, Probabilities and Knowledge

Mathematical logic can be studies at many different levels. In this book I will concentrate on propositional logic, and I will look at propositions as they are formulated in ordinary, everyday language as primitive entities. For a more formal approach to propositional logic including axioms and a separation between syntax and semantics, see for instance (Walicki 2012).

Propositions A and B can be connected: $A \vee B$ means that A or B is true, while $A \wedge B$ means that both A and B are true, similarly for the connection between more propositions. Also, $\neg A$ means that A is not true. We let $\perp$ denote an impossible proposition, while $\top$ denotes a proposition which is always true. In ordinary texts in mathematical logic one usually works with a finite number of propositions A_i. I will allow for an infinite, even uncountable number of propositions, so that propositions of the form 'The rain tomorrow will amount to less than or equal to x mm' will be permitted for different x.

There is a close connection between propositional logic and set theory. The translation is straightforward: $\vee$ translates into $\cup$, while $\wedge$ translates into $\cap$; $\neg A$ corresponds to A^c, while $\perp, \top$ correspond to $\emptyset, \Omega$, assuming that all the sets are subsets of Ω.

One can also define probabilities of propositions; in fact this is often done in elementary probability texts. With the above translations, there is a close connection to Kolmogorov's axioms; see Sect. 2.1. For instance $P(A_1 \vee A_2 \vee \ldots) = P(A_1) + P(A_2) + \ldots$ if the A_i's satisfy $A_i \wedge A_j = \perp$ for each pair. Also $P(\neg A) = 1 - P(A)$. The rule $P(A \vee B) = P(A) + P(B) - P(A \wedge B)$ is always true. It can be proved rigorously, but it can also be motivated by a Venn diagram from the analogue with set theory.

Conditional probabilities can be defined by $P(A|B) = P(A \wedge B)/P(B)$ when $P(B) > 0$. In this book I need the more general notion of conditional probability given a σ-algebra of propositions $\mathcal{B}$, and then it seems like we may need to assume a little more structure. Assume thus that there exists a countable collection of atomic propositions $\{C_i\}$ such that all other propositions A can be formed by combining the C_i's by $\vee$'s, such that $C_i \wedge C_j = \perp$ for pairs and $\top = \bigvee_i C_i$. This assumption simplifies the discussion. It is satisfied in the case of a finite number of propositions closed under $\wedge$ and $\neg$. In general we can think of the C_i's as formed by combining all propositions of interest through $\wedge$'s. The whole σ-algebra $\mathcal{F}$ is generated by the C_i's.

Let now the sub-σ-algebra $\mathcal{B}$ be generated by $\{B_j\}$, partly a subset of $\{C_i\}$ and partly formed by taking $\vee$ of some C_i, such that $B_j \wedge B_k = \perp$ for pairs and such that $\top = \bigvee_j B_j$. Then we can define

$$P(A|\mathcal{B}) = \sum_j P(A|B_j)\mathbf{1}(B_j),$$

where $\mathbf{1}(B_j) = 1$ if B_j is true, 0 if it is not true. From this, $P(A|\mathcal{B})$ is uniquely defined except on a set with probability 0. The analogue of the Radon-Nikodym definition (2.1) is then

$$\int_B P(A|\mathcal{B})dP = P(A \wedge B)$$

for all $B \in \mathcal{B}$. One of the open questions of this book is whether this formula can be generalized in the context of propositional logic, and then can be taken as a general definition of $P(A|\mathcal{B})$.

In the probabilistic treatment I assumed in Sect. 3.1 that the observations and the parameters could be defined on the same underlying probability space. In the present setting I assume that all statements regarding conceptual variables can be given as compatible propositions. The concept of an epistemic process is central in this book. Before any observations are made, all statements of the form $\theta = u$ are unknown, where θ is the relevant epistemic variable. After the observations are done, some proposition $A_k : (\theta = u_k)$ is known to some agent i in the simplest case. The statement that A_k is known to agent i may be written $K_i A_k$, and the statement that agent j knows that i knows A_k may be written $K_j K_i A_k$. A survey of the formal propositional logic related to such statements is given by Halpern (1995).

References

Barut, A. S., & Raczka, R. (1985). *Theory of group representation and applications*. Warsaw: Polish Scientific Publishers.

Halpern, J. Y. (1995). Reasoning about knowledge: A survey. In D. M. Gabbay, C. J. Hogger, & J. A. Robinson, J. A. (Eds.), *Handbook of logic in artificial intelligence and logic programming*. (Vol. 4). *Epistemic and temporal reasoning*. Oxford: Oxford University Press.

Helland, I. S. (2010). *Steps towards a unified basis for scientific models and methods*. Singapore: World Scientific.

Walicki, M. (2012). *Introduction to mathematical logic*. Hoboken, NJ: World Scientific.

Wijsman, R. A. (1990). *Invariant measures on groups and their use in statistics. Lecture notes - Monograph series* (Vol. 14). Hayward, CA: Institute of Mathematical Statistics.

Bibliography

Aerts, D., & Gabora, L. (2005a). A theory of concepts and their properties I. The structure of sets of contexts and properties. *Kybernetes, 34*, 167–191.

Aerts, D., & Gabora, L. (2005b). A theory of concepts and their properties II. A Hilbert space representation. *Kybernetes, 34*, 192–221.

Aerts, D., Sozzo, S., & Tapia, J. (2014). Identifying quantum structures in the Ellsberg paradox. *International Journal of Theoretical Physics, 53*, 3666–3682.

Ashitani, M., & Azgomi, M. A. (2015). A survey of quantum-like approaches to decision making and cognition. *Mathematical Social Sciences, 75*, 49–80.

Bagarello, F. (2013). *Quantum dynamics for classical systems.* Hobroken, NJ: Wiley.

Ballentine, L. E. (1998). *Quantum mechanics. A modern development.* Singapore: World Scientific.

Bargmann, V. (1964). Note on Wigner's theorem on symmetry operations. *Journal of Mathematical Physics, 5*, 862–868.

Barndorff-Nielsen, O. E., Gill, R. D., & Jupp, P. E. (2003). On quantum statistical inference. *Journal of the Royal Statistical Society B, 65*, 775–816.

Barut, A. S., & Raczka, R. (1985). *Theory of group representation and applications.* Warsaw: Polish Scientific Publishers.

Bell, J. S. (1975). The theory of local beables. Reprinted in Bell (1987).

Bell, J. S. (1987). *Speakable and Unspeakable in quantum mechanics.* Cambridge: Cambridge University Press.

Berger, J. O., & Wolpert, R. L. (1988). *The likelihood principle.* Hayward, CA: Institute of Mathematical Statistics.

Bernardo, J. M., & Smith, A. F. M. (1994). *Bayesian Theory.* Chichester: Wiley.

Bickel, P. J., & Doksum, K. A. (2001). *Mathematical statistics. Basic ideas and selected topics.* (2nd ed.). Upper Saddle River, NJ: Prentice Hall.

Bing-Ren, L. (1992). *Introduction to operator algebras.* Singapore: World Scientific.

Birnbaum, A. (1962). On the foundation of statistical inference. *Journal of the American Statistical Association, 57*, 269–326.

Bjørnstad, J. F. (1990). Predictive likelihood: A review. *Statistical Science, 5*, 242–265.

Bohr, N. (1935a). Quantum mechanics and physical reality. *Nature, 136*, 65.

Bohr, N. (1935b). Can quantum-mechanical description of physical reality be considered complete? *Physical Review, 48*, 696–702.

Box, G. E. P., & Tiao, G. C. (1973). *Bayesian Inference in Statistical Analysis.* New York: Wiley.

Breiman, L. (2001). Statistical modeling: The two cultures. *Statistical Science, 16*, 199–231.

Briggs, G. A. D., Butterfield, J. N., & Zeilinger, A. (2013). The Oxford questions on the foundation of quantum physics. *Proceedings of the Royal Society A, 469*, 20130299.

© Springer-Verlag GmbH Germany, part of Springer Nature 2018

I. S. Helland, *Epistemic Processes*, https://doi.org/10.1007/978-3-319-95068-6

Brody, T. (1993). In L. de la Pera & P. Hodgson (Eds.), *The philosophy behind physics*. Berlin: Springer.

Brown, L. M. (Ed.) (2005). *Feynman's thesis. A new approach to quantum theory*. New Jersey: World Scientific.

Busch, P. (2003). Quantum states and generalized observables: A simple proof of Gleason's Theorem. *Physical Review Letters, 91*(12), 120403.

Busch, P., Lahti, P. J., & Mittelstaedt, P. (1991). *The quantum theory of measurement*. Berlin: Springer.

Busch, P., Lahti, P., Pellonpää, J.-P., & Ylinen, K. (2016). *Quantum measurement*. Berlin: Springer.

Busemeyer, J. R., & Bruza, P. (2012). *Quantum models of cognition and decision*. Cambridge: Cambridge University Press.

Cabello, A. (2015). Interpretations of quantum theory: A map of madness. arXiv: 1509.0471v1 [quant-ph]

Casella, G., & Berger, R. L. (1990). *Statistical inference*. Pacific Grove, CA: Wadsworth and Brooks.

Casinelli, G., & Lahti, P. (2016). An axiomatic basis for quantum mechanics. *Foundations of Physics, 46*, 1341–1373.

Caves, C. M., Fuchs, C. A., & Schack, R. (2002). Quantum probabilities as Bayesian probabilities. *Physical Review, A65*, 022305.

Cetina, K. K. (1999). *Epistemic cultures. How the sciences make knowledge*. Cambridge, MA: Harvard University Press.

Charrakh, O. (2017). On the reality of the wavefunction. arXiv: 1706.01819 [physics.hist-ph]

Chiribella, G., D'Ariano, G. M., & Perinotti, P. (2010). Informational derivation of quantum theory. arXiv: 1011.6451 [quant-ph]

Cochran, W. G. (1977). *Sampling techniques*. (3rd ed.). New York: Wiley.

Colbeck, R., & Renner, R. (2013). A short note on the concept of free choice. arXiv: 1302.4446 [quant-ph]

Congdon, P. (2006). *Bayesian statistical modelling* (2nd ed.). Chichester: Wiley.

Conway, J., & Kochen, S. (2006). The free will theorem. *Foundations of Physics, 36*, 1441–1473.

Conway, J., & Kochen, S. (2008). The strong free will theorem. arXiv: 0807.3286 [quant-ph]

Cook, R. D. (2007). Fisher lecture: Dimension reduction in regression. *Statistical Science, 22*, 1–26.

Cook, R. D., Helland, I. S., & Su, Z. (2013). Envelopes and partial least squares regression. *Journal of the Royal Statistical Society Series B, 75*, 851–877.

Cook, R. D., Li, B., & Chiaromonte, F. (2010). Envelope models for parsimonious and efficient multivariate linear regression. *Statistica Sinica, 20*, 927–1010.

Cox, D. R. (1958). Some problems connected with statistical inference. *Annals of Statistics, 29*, 357–372.

Cox, D. R. (1971). The choice between ancillary statistics. *Journal of the Royal Statistical Society. Series B, 33*, 251–255.

Cox, D. R. (2006). *Principles of statistical inference*. Cambridge: Cambridge University Press.

Cox, D. R., & Donnelly, C. A. (2011). *Principles of applied statistics*. Cambridge: Cambridge University Press.

Eaton, M. L. (1989). *Group invariance in applications in statistics*. Hayward, CA: Institute of Mathematical Statistics and American Statistical Association.

Efron, B. (1998). R.A. Fisher in the 21st century. *Statistical Science, 13*, 95–122.

Efron, B. (2015). Frequency accuracy of Bayesian estimates. *Journal of the Royal Statistical Society B, 77*, 617–646.

Eichberger, J., & Pirner, H. J. (2017). Decision theory with a Hilbert space as a probability space. arXiv: 1707.07556 [quant-ph].

Einstein, A., Podolsky, B., & Rosen, N. (1935). Can quantum-mechanical description of physical reality be considered complete? *Physical Review, 47*, 777–780.

Everett, H. III (1973). The theory of the universal wave function. In N. Graham, B. DeWitt (Eds.), *The many worlds interpretation of quantum mechanics*. Princeton: Princeton University Press.

Feynman, R. P. (1985). *QED. The strange theory of light and matter.* Princeton: Princeton University Press.

Fields, C. (2011). Quantum mechanics from five physical assumptions. arXiv: 1102.0740 [quant-ph].

Fisher, R. A. (1922). On the mathematical foundations of theoretical statistics. *Philosophical Transactions of the Royal Society of London. Series A, 222,* 309–368. Reprinted in: Fisher R. A. Contribution to Mathematical Statistics. Wiley, New York (1950)

Fivel, D. I. (2012). Derivation of the rules of quantum mechanics from information-theoretic axioms. *Foundations of Physics, 42,* 291–318.

Frieden, B. R. (1998). *Physics from Fisher Information. A Unification.* Cambridge: Cambridge University Press.

Frieden, B. R. (2004). *Science from Fisher Information. A Unification.* Cambridge: Cambridge University Press.

Fuchs, C. A. (2002). Quantum mechanics as quantum information (and only a little more). In Khrennikov, A. (Ed.), *Quantum theory: Reconsideration of Foundations.* Växjö: Växjö University Press.

Fuchs, C. A. (2010). QBism, the perimeter of quantum Bayesianism. arXiv: 1003.5209v1 [quant-ph].

Fuchs, C. A. (2016). On participatory realism. arXiv: 1601.04360v2 [quant-ph].

Fuchs, C. A., Mermin, N. D., & Schack, R. (2013). An introduction to QBism with an application to the locality of quantum mechanics. arXiv: 1311.5253v1 [quant-ph].

Fuchs, C. A., & Peres, A. (2000). Quantum theory needs no interpretation. *Physics Today, S-0031-9228-0003-230-0;* Discussion: *Physics Today, S-0031-9228-0009-220-6.*

Fuchs, C. A., & Schack, R. (2011). A quantum-Bayesian route to quantum-state space. *Foundations of Physics, 41,* 345–356.

Gelman, A., & Robert, C. P. (2013). "Not only defended but also applied": The perceived absurdity of Bayesian inference. *The American Statistician, 67,* 1–5.

Gill, R., Guta M., & Nussbaum, M. (2014). New horizons in statistical decision theory. *Mathematisches Forschungsinstitut Oberwolfach.* Report No. 41.

Giulini, D. (2009). Superselection rules. arXiv: 0710.1516v2 [quant-ph].

Griffiths, R. B. (2014). The consistent history approach to quantum mechanics. In E. N. Zalta (Ed.), *Stanford encyclopedia of philosophy.* Stanford: Metaphysics Research Lab, Stanford University.

Griffiths, R. B. (2017a). What quantum measurements measure. *Physical Review A, 96,* 032110.

Griffiths, R. B. (2017b). Quantum information: What is it all about? *Entropy, 19,* 645.

Hall, M. J. W. (2011). Generalizations of the recent Pusey-Barrett-Rudolph theorem for statistical models of quantum phenomena. arXiv: 1111.6304 [quant-ph].

Halpern, J. Y. (1995). Reasoning about knowledge: A survey. In: D. M. Gabbay, C. J. Hogger, & J. A. Robinson (Eds.), *Handbook of logic in artificial intelligence and logic programming* (Vol. 4). *Epistemic and temporal reasoning.* Oxford: Oxford University Press.

Hammond, P. J. (2011). *Laboratory games and quantum behavior. The normal form with a separable state space.* Working paper. Department of Economics, University of Warwick.

Hardy, L. (2001). Quantum theory from five reasonable axioms. arXiv: 0101012v4 [quant-ph].

Hardy, L. (2011). Reformulating and reconstructing quantum theory. arXiv: 1104.2066v1 [quant-ph].

Hardy, L. (2012a). Are quantum states real? arXiv: 1205.1439 [quant-ph].

Hardy, L. (2012b). The operator tensor formulation of quantum theory. arXiv: 1201.4390v1 [quant-ph].

Hardy, L. (2013). Reconstructing quantum theory. arXiv: 1303.1538v1 [quant-ph].

Hardy, L., & Spekkens, R. (2010). Why physics needs quantum foundations. arXiv: 1003.5008 [quant-ph].

Harris, B. (1982). Entropy. In S. Kotz & N. L. Johnson . *Encyclopedia of statistical sciences.* Hoboken, NJ: Wiley.

Hastie, T., Tibshirani, R., & Friedman, J. (2009). *The elements of statistical learning. Data mining, inference, and prediction.* Springer series in statistics.

Haven, E., & Khrennikov, A. (2013). *Quantum social science.* Cambridge: Cambridge University Press.

Haven, E., & Khennikov, A. (2016). Quantum probability and mathematical modelling of decision making. *Philosophical Transactions of the Royal Society A, 374,* 20150105.

Helland, I. S. (1990). Partial least squares regression and statistical models. *Scandinavian Journal of Statistics, 17,* 97–114.

Helland, I. S. (1995). Simple counterexamples against the conditionality principle. *The American Statistician, 49,* 351–356. *Discussion, 50,* 382–386.

Helland, I. S. (2004). Statistical inference under symmetry. *International Statistical Review, 72,* 409–422.

Helland, I. S. (2006). Extended statistical modeling under symmetry; the link toward quantum mechanics. *Annals of Statistics, 34,* 42–77.

Helland, I. S. (2008). Quantum mechanics from focusing and symmetry. *Foundations of Physics, 38,* 818–842.

Helland, I. S. (2010). *Steps Towards a Unified Basis for Scientific Models and Methods.* Singapore: World Scientific.

Helland, I. S. (2017). The conception of God as seen from research on the foundation of quantum mechanics. *Dialogo Journal, 4*(1), 259–267.

Helland, I. S., Sæbø, S., Almøy, T., & Rimal, R. (2018). Model and estimators for partial least squares. *Journal of Chemometrics* (to appear).

Helland, I. S., Sæbø, S., & Tjelmeland, H. (2012). Near optimal prediction from relevant components. *Scandinavian Journal of Statistics, 39,* 695–713.

Helstrom, C. W. (1976). *Quantum detection and estimation theory.* New York: Academic.

Hermansen, G., Cunen, C., & Stoltenberg, E. A. (2017). Ny bok: Confidence, likelihood, probability. statistical inference with confidence distributions. (In Norwegian). *Tilfeldig Gang, 34*(1), 9–14.

Holevo, A. S. (1982). *Probabilistic and statistical aspects of quantum theory.* Amsterdam: North-Holland.

Holevo, A. S. (2001). *Statistical structure of quantum theory.* Berlin: Springer.

Jalger, G. (2018). Developments in quantum probability and the Copenhagen approach. *Entropy, 20,* 420–438.

Kass, R. E., & Wasserman, L. (1996). The selection of prior distributions by formal rules. *Journal of the American Statistical Association, 91,* 1343–1370.

Khrennikov, A. (2010). *Ubiquitous quantum structure.* Berlin: Springer.

Khrennikov, A. (2014). *Beyond quantum.* Danvers, MA: Pan Stanford Publishing.

Khrennikov, A. (2016a). Quantum Bayesianism as a basis of general theory of decision making. *Philosophical Transactions of the Royal Society A, 374,* 20150245.

Khrennikov, A. (2016b). After Bell. arXiv: 1603.086774 [quant-ph].

Klebaner, F. C. (1998). *Introduction to stochastic calculus with applications.* London: Imperial College Press.

Knapp, A. W. (1986). *Representation theory of semisimple groups.* Princeton, NJ: Princeton University Press.

Kochen, S., & Specker, E. P. (1967). The problem of hidden variables in quantum mechanics. *Journal of Mathematics and Mechanics, 17,* 59–87.

Kuhlmann, M. (2013). What is real? *Scientific American, 309*(2), 32–39.

LeCam, L. (1990). Maximum likelihood: an introduction. *International Statistical Review, 58,* 153–171.

Lehmann, E. L. (1999). *Elements of large-sample theory.* New York: Springer.

Lehmann, E. L., & Casella, G. (1998). *Theory of point estimation.* New York: Springer.

Leifer M. F. (2014). Is the quantum state real? An extended review of ψ-ontology theorems. arXiv.1409.1570v2 [quant-ph].

Ma, Z.-Q. (2007). *Group theory for physicists.* Hoboken, NJ: World Scientific.

Martens, H., & Næs, T. (1989). *Multivariate calibration*. Hoboken, NJ: Wiley.

Masanes, L. (2010). Quantum theory from four requirements. arXiv: 1004.1483 [quant-ph].

McCullagh, P. (2002). What is a statistical model? *Annals of Statistics, 30*, 1225–1310.

McCullagh, P., & Han, H. (2011). On Bayes's theorem for improper mixtures. *Annals of Statistics, 39*, 2007–2020.

Mermin, N. D. (1985). Is the moon there when nobody looks? *Physics Today, 38*, 38–47.

Mermin, N. D. (2014). Why QBism is not the Copenhagen interpretation and what John Bell might have thought of it. arXiv.1409.2454 [quant-ph].

Messiah, A. (1969). *Quantum mechanics* (Vol. II). Amsterdam: North-Holland.

Mirman, R. (1995). *Group theoretical foundations of quantum mechanics*. Lincoln, NE: iUniverse.

Murphy, G. J. (1990). *C*-algebras and operator theory*. Boston: Academic.

Næs, T., & Helland, I. S. (1993). Relevant components in regression. *Scandinavian Journal of Statistics, 20*, 239–250.

Nelson, E. (1967). *Dynamical theories of Brownian motion*. Princeton: Princeton University Press.

Neyman, J., & Scott, E. L. (1948). Consistent estimators based on partially consistent observations. *Econometrica, 16*, 1–16.

Nisticò, G., & Sestito, A. (2011). Quantum mechanics, can it be consistent with locality? *Foundations of Physics, 41*, 1263–1278.

Norsen, T., & Nelson, S. (2013). Yet another snapshot of fundamental attitudes toward quantum mechanic. arXiv:1306.4646v2 [quant-ph].

Östborn, P. (2016). A strict epistemic approach to physics. arXiv:1601.00680v2 [quant-ph].

Östborn, P. (2017). Quantum mechanics from an epistemic state space. arXiv:1703.08543 [quant-ph].

Pearl, J. (2009). *Causality. Models, reasoning and inference*. (2nd ed.). Cambridge: Cambridge University Press.

Penrose, R. (2016). *Fashion, faith and fantasy in the new physics of the universe*. Princeton, NJ: Princeton Universiy Press.

Peres, A. (1993). *Quantum theory: concepts and methods*. Dordrecht: Kluwer.

Plotnitsky, A. (2013). *Niels Bohr and complementarity. An introduction*. New York: Springer.

Pothos, E. M., & Busemeyer, J. R. (2013). Can quantum probability provide a new direction for cognitive modeling? With discussion. *Behavioral and Brain Sciences, 36*, 255–327.

Pusey, M. F., Barrett, J., & Rudolph, T. (2012). On the reality of quantum states. *Nature Physics, 8*, 475–478.

Reid, N. (1995). The roles of conditioning in inference. Statistical Science, 10(2), 138-157.

Robinson, P. M. (1991). Consistent nonparametric entropy-based testing. *Review of Economic Studies, 58*, 437–453.

Rovelli, C. (2016). An argument against a realistic interpretation of the wave function. *Foundations of Physics, 46*, 1229–1237.

Sæbø, S., Almøy, T. and Helland, I. S. (2015). simrel - a versatile tool for linear model data simulation based on the concept of a relevant subspace and relevant predictors. *Chemometrics and Intelligent Laboratory Systems, 146*, 128–135.

Schack, R. (2006). Bayesian probability in quantum mechanics. In *Proceedings of Valencia/ ISBA World Meeting on Bayesian Statistics*.

Schlosshauer, M. (2007). *Decoherence and the quantum-to-classical transition*. New York: Springer.

Schlosshauer, M., Kofler, J., & Zeilinger, A. (2013). A snapshot of fundamental attitudes toward quantum mechanics. *Studies in History and Philosophy of Modern Physics, 44*, 222–238.

Schweder, T., & Hjort, N. L. (2002). Confidence and likelihood. *Scandinavian Journal of Statistics, 29*, 309–332.

Schweder, T., & Hjort, N. L. (2016). *Confidence, likelihood, probability. Statistical inference with confidence distributions*. Cambridge: Cambridge University Press.

Searle, S. R. (1971). *Linear models*. New York: Wiley.

Sen, P. K., & Singer, J. M. (1993). *Large sample methods in statistics*. London: Chapman and Hall.

Shannon, C. E., & Weaver, W. (1949). *The mathematical theory of communication.* Champaign, IL: University of Illinois Press.

Smilga, W. (2017). Towards a constructive foundation of quantum mechanics. *Foundations of Physics, 47*, 149–159.

Smolin, L. (2011). A real ensemble interpretation of quantum mechanics. aXiv. 1104.2822 [quant-ph].

Sornette, D. (2014). Physics and financial economics (1776–2014): puzzles, Ising and agent-based models. *Reports on Progress in Physics, 77*, 062001.

Spekkens, R. W. (2007). In defense of the epistemic view of quantum states: a toy theory. *Phys. Rev. A, 75*, 032110.

Spekkens R. W. (2014). Quasi-quantization: classical statistical theories with an epistemic restriction. arXiv.1409.304 [quant-ph]

Stigler, S. M. (1976). Discussion of "On rereading R.A. Fisher" by L.J. Savage. *Annals of Statistics, 4*, 498–500.

Tammaro, E. (2014). Why current interpretations of quantum mechanics are deficient. arXiv1408.2083v2 [quant-ph].

Taraldsen, G., & Lindqvist, B. H. (2010). Improper priors are not improper. *The American Statistician, 64*, 154–158.

Timpson, C. G. (2008). Quantum Bayesianism: A study. *Studies in History an Philosophy of Modern Physics, 39*, 579–609.

Tversky, A., & Kahneman, D. (1983). Extensional versus intuitive reasoning: The cojunction fallacy in probability judgements. *Psychological Review, 90*, 293–315.

Vedral, V. (2011). Living in a quantum world. *Scientific American, 304*(6), 20–25.

Venema, Y. (2001). Temporal logic. In: Goble, L. (Ed.), *The Blackwell guide to philosophical logic.* Hoboken, NJ.: Blackwell.

von Baeyer, H. C. (2013). Quantum weirdness? It's all in your mind. *Scientific American, 308*(6), 38–43.

von Baeyer, H. C. (2016). *QBism: The future of quantum physics.* Harvard: Harvard University Press.

von Neumann, J. (1927). Wahrscheinlichkeitstheoretischer Aufbau der Quantenmechanik. *Nachrichten von der Gesellschaft der Wissenschaften zu Göttingen, Mathematisch-Physikalische Klasse, 1927*, 245–272.

von Neumann, J. (1932). *Mathematische Grundlagen der Quantenmechanik.* Berlin: Springer.

Walicki, M. (2012). *Introduction to Mathematical Logic.* New Jersey: World Scientific.

Wasserman, L. (2004). *All of statistics. A concise course in statistical inference.* New York: Springer.

Wetterich, C. (2008a). Probabilistic observables, conditional correlations, and quantum physics. arXiv: 0810.0985v1 [quant-ph].

Wetterich, C. (2008b). Quantum entanglement and interference from classical statistics. arXiv: 0809.2671v1 [quant-ph].

Wetterich, C. (2009). Zwitters: Particles between quantum and classical. arXiv: 0911.1261v2 [quant-ph].

Wetterich, C. (2010a). Quantum particles from coarse grained classical particles in phase space. arXiv: 1003.3351v1 [quant-ph].

Wetterich, C. (2010b). Probabilistic time. arXiv: 1002.2593v1 [quant-ph].

Wetterich, C. (2010c). Quantum particles from classical probabilities in phase space. arXiv: 10003.0772v1 [quant-ph].

Wetterich, C. (2010d). Quantum mechanics from classical statistics. *Annals of Physics, 325*, 852–884.

Wigner, E. (1939). On unitary representations of the inhomogeneous Lorentz group. *Annals of Mathematics, 40*, 149–204.

Wigner, E. P. (1959). *Group theory and its application to the quantum mechanics of atomic spectra.* New York: Academic.

Wijsman, R. A. (1990). *Invariant measures on groups and their use in statistics. Lecture Notes - Monograph Series* (Vol. 14). Hayward, CA: Institute of Mathematical Statistics.

Wootters, W. K. (1980). *The acquisition of information from quantum measurements.* PhD Thesis. Center for Theoretical Physics. The University of Texas at Austin.

Wootters, W. K. (2004). Quantum measurements and finite geometry. arXiv:quant-ph/0406032v3.

Xie, M., & Singh, K. (2013). Confidence distributions, the frequentist distribution estimator of a parameter - a review. Including discussion. *International Statistical Review, 81*, 1–77.

Yukalov, V. I., & Sornette D. (2008). Quantum decision theory as a quantum theory of measurement. *Physics Letters A, 372*, 6867–6871.

Yukalov, V. I., & Sornette, D. (2009). Processing information in quantum decision theory. *Entropy, 11*, 1073–1120.

Yukalov, V. I., & Sornette, D. (2010). Mathematical structure of quantum decision theory. *Adv. Compl. Syst., 13*, 659–698.

Yukalov, V. I., & Sornette, D.(2011). Decision theory with prospect interference and entanglement. *Theory and Decision, 70*, 383–328.

Yukalov, V. I., & Sornette, D. (2014). How brains make decisions. *Springer Proceedings in Physics, 150*, 37–53.

Yukalov, V. I., Yukalova, E. P., & Sornette, D. (2017). Information processing by networks of quantum decision makers. arXiv: 1712.05734 [physics.soc-ph].

Zeilinger, A. (1999). A foundational principle for quantum mechanics. *Foundations of Physics, 29*, 631–643.

Zeilinger, A. (2010). *Dance of the photons. From Einstein to quantum teleportation.* New York: Farrar, Straus and Giroux.

Index

© Springer-Verlag GmbH Germany, part of Springer Nature 2018
I. S. Helland, *Epistemic Processes*, https://doi.org/10.1007/978-3-319-95068-6

Printed in the United States
By Bookmasters